SCHAUM'S *Easy* OUTLINES

LOGIC

Other Books in Schaum's Easy Outline Series Include:

SCHAUM'S *Easy* OUTLINES

LOGIC

BASED ON SCHAUM'S
Outline of Theory and Problems of Logic
BY
JOHN NOLT, Ph.D.
DENNIS ROHATYN, Ph.D.
ACHILLE VARZI, Ph.D.

ABRIDGEMENT EDITOR:
ALEX M. MCALLISTER, Ph.D.

SCHAUM'S OUTLINE SERIES
McGRAW-HILL

New York   Chicago   San Francisco   Lisbon   London   Madrid   Mexico City
Milan   New Delhi   San Juan   Seoul   Singapore   Sydney   Toronto

JOHN NOLT is Associate Professor of Philosophy at the University of Tennessee, Knoxville, where he has taught since receiving his doctorate from the Ohio State University in 1978. He is the author of *Informal Logic: Possible Worlds and Imagination* and numerous articles on logic, metaphysics, and the philosophy of mathematics.

DENNIS ROHATYN is Professor of Philosophy at the University of San Diego, where he has taught since 1977. He is the author of *Two Dogmas of Philosophy*, *The Reluctant Naturalist*, and many other works. He is a regular symposiast on critical thinking at national and regional conferences. In 1987, he founded the Society for Orwellian Studies.

ACHILLE VARZI is Associate Professor of Philosophy at Columbia University, New York. His books include *Holes and Other Superficialities* and *Parts and Places* (both with R. Casati) and *An Essay in Universal Semantics*. He is also the author of numerous articles on logic, metaphysics, and the philosophy of language and an editor of *The Journal of Philosophy*.

ALEX M. McALLISTER is Associate Professor of Mathematics at Centre College in Danville, Kentucky. He received a B.S. degree from Virginia Polytechnic Institute and State University and a Ph.D. from the University of Notre Dame. His scholarly interests include mathematical logic and foundations and computability theory. His articles have been published in *Archive for Mathematical Logic* and *Journal of Symbolic Logic*.

1 2 3 4 5 6 7 8 9 DOC/DOC 0 9 8 7 6 5

ISBN 0-07-145535-3

Contents

Chapter 1
ARGUMENT STRUCTURE

IN THIS CHAPTER:

- ✔ *What Is an Argument?*
- ✔ *Identifying Arguments*
- ✔ *Complex Arguments*
- ✔ *Argument Diagrams*
- ✔ *Convergent Arguments*
- ✔ *Implicit Statements*
- ✔ *Use and Mention*
- ✔ *Formal vs. Informal Logic*

What is an Argument?

Logic is the study of arguments. An *argument* is a sequence of statements of which one is intended as a *conclusion* and the others, the *premises*, are intended to prove or at least provide some evidence for the conclusion. Here is an example:

All humans are mortal. Socrates is human.
Therefore, Socrates is mortal.

In this argument, the first two statements are premises intended to prove the conclusion that Socrates is mortal.

1

The premises and the conclusion of an argument are always *statements* or *propositions*, as opposed to questions, commands, or exclamations. A statement is an assertion that is either true or false and is typically expressed by a declarative sentence. Here are some examples:

> Dogs do not fly.
> Snow is red or blue.
> My brother is an entomologist.

The first sentence expresses a statement that is true, the second expresses a false statement, and the last is true or false depending on whether or not the brother of the speaker is in fact an entomologist. In contrast, the following are not statements:

> Who is the author of *The Man Without Qualities*?
> Please do not call after 11 p.m.

Non-statements, such as questions, commands, or exclamations, are neither true nor false.

Example 1.1. One of the following is an argument. Identify its premises and conclusions.

(a) He's a Leo, since he was born in the first week of August.
(b) Is there anyone here who understands this document?
(c) John went home. However, Mary went to the movie.

(a) Premise: He was born in the first week of August.
    Conclusion: He's a Leo.
(b) Not an argument; this is just a question.
(c) Not an argument; these are just two unrelated sentences.

Notice that the conclusion of the argument given in Example 1.1 is at the beginning of the sentence. The conclusion may occur anywhere in the argument, but the beginning and end are the most common positions.

Remember

For purposes of analysis, it is customary to list the premises first, each on a separate line, and then to give the conclusion.

The conclusion is often marked by the symbol ∴, which means "therefore." This format is called *standard form*. Thus, the standard form of our initial example is:

> All humans are mortal.
> Socrates is human.
> ∴ Socrates is mortal.

Identifying Arguments

Argument occurs only when someone intends a set of premises to support or prove a conclusion. This intention is often expressed by the use of *inference indicators*, which are words or phrases used to signal the presence of an argument. They are of two kinds: *conclusion indicators* and *premise indicators*. Here are some typical examples:

Conclusion Indicators	Premise Indicators
Therefore	For
Thus	Since
Hence	Because
So	Assuming that

Premise and conclusion indicators are the main clues in identifying arguments and analyzing their structure. Consider the argument:

> He is not at home, so he has gone to the movie.

The conclusion indicator "so" signals that "He is not at home" is a conclusion supported by the premise "He has gone to the movie". To contrast, consider the similar argument:

He is not at home, since he has gone to the movie.

Here the premise indicator "since" indicates that "He has gone to the movie" is a premise supporting the conclusion "He is not at home".

Example 1.2. Use the inference indicators of an argument below to determine its inferential structure, and then write it in standard form.

[1][Gold-argon compounds are not likely to be produced even in the laboratory, much less in nature,] **since** [2][it is difficult to make argon react with anything,] and **since** [3][gold, too, forms few compounds.]

We have bold-faced the inference indicators for emphasis and bracketed and numbered each statement for ease of reference. The argument consists of a compound sentence whose three component sentences are linked together by two occurrences of the premise indicator "since", each of which introduce a premise in statements 2 and 3. The conclusion is statement 1. In standard form the argument is:

It is difficult to make argon react with anything.
Gold, too, forms few compounds.
∴ Gold-argon compounds are not likely to be produced even in the laboratory, much less in nature.

Expressions that function in some contexts as inference indicators generally have other functions in other contexts. Thus, the notion of an inference indicator should not be taken too rigidly. For example, the word "since" in the following sentence indicates duration and does not function as a premise indicator.

It has been six years since we went to France.

Note!

Some arguments have no indicators at all. In such cases, we must rely on contextual clues or our understanding of the author's intentions in order to differentiate premises from conclusions.

Example 1.3. Rewrite the following argument in standard form.

> Some politicians are hypocrites. They say we should pay more taxes if the national deficit is to be kept under control. But then they waste huge amounts of money on their election campaigns.

The author's intention is to establish that some politicians are hypocrites. In standard form the argument is:

> Some politicians say we should pay more taxes if the national deficit is to be kept under control.
> They waste huge amounts of money on their election campaigns.
> ∴ They are hypocrites.

Complex Arguments

Some arguments proceed in stages. First a conclusion is drawn from a set of premises; then, that conclusion (perhaps in conjunction with some other statements) is used as a premise to draw a further conclusion, which may function as a premise for yet another conclusion, and so on. Such a structure is called a complex argument. For example, the following argument is complex:

> All rational numbers are expressible as a ratio of integers.
> Pi is not expressible as a ratio of integers.
> ∴ Pi is not a rational number.
> Pi is a number.
> ∴ There exists at least one non-rational number.

Each of the simple steps of reasoning that are linked together to form a complex argument is an argument in its own right. The complex argument consists of two such steps. The first three statements make up the first argument, and the second three statements make up the second argument. The third statement, "Pi is not a rational number", is a component of both steps, functioning as the conclusion of the first and a premise of the second.

Argument Diagrams

Argument diagrams are a convenient way of representing inferential structure. To diagram an argument, identify the inference indicators and bracket and number each statement. If several premises function together in a single step of reasoning, write their numbers in a horizontal row, joined by plus signs, and underline this row of numbers. If a step of reasoning has only one premise, simply write its number. In either case, draw an arrow downward from the number(s) representing the premise(s) to the number representing the conclusion of the step. Repeat this procedure for complex arguments.

Example 1.4. Diagram the following argument.

> (1)[Today is either Tuesday or Wednesday.] But (2)[it can't be Wednesday,] **since** (3)[the doctor's office was open this morning,] and (4)[that office is always closed on Wednesday.] **Therefore**, (5)[today must be Tuesday.]

The premise indicator "since" signals that statements 3 and 4 are premises supporting statement 2. The conclusion indicator "therefore" signals that statement 5 is a conclusion from previously stated premises. Consideration of the context and meaning of each sentence reveals that the premises directly supporting statement 5 are statements 1 and 2. This the argument should be diagrammed as:

$$\frac{3+4}{}$$
$$\downarrow$$
$$\frac{1+2}{}$$
$$\downarrow$$
$$5$$

You Need to Know

The plus signs in the diagram mean "together with" or "in conjunction with," and the arrows mean "is intended as evidence for."

Thus the meaning of the diagram of Example 1.4 is: "3 together with 4 is intended as evidence for 2, which together with 1 is intended as evidence for 5."

Example 1.5. Diagram the following argument.

(1)[The check is void unless it is cashed within 30 days.] (2)[The date on the check is September 2,] and (3)[it is now October 8.] **Therefore,** (4)[the check is now void.] (5)[You cannot cash a check which is void.] **So** (6)[you cannot cash this one.]

$$1 + 2 + 3$$
$$\downarrow$$
$$4 + 5$$
$$\downarrow$$
$$6$$

Notice that premise 1 is treated as a single unit; the word "unless" is not an inference indicator but a connective. That is, a locution that joins two sentences to form a compound statement.

Convergent Arguments

If an argument contains several steps of reasoning that all support the same (final or intermediate) conclusion, the argument is said to be *convergent*. Consider the following:

One should quit smoking. It is very unhealthy, and it is annoying to the bystanders.

Here the statements "smoking is unhealthy" and "smoking is annoying" function as independent reasons for the conclusion that "one should quit smoking". We do not, for example, need to assume the first premise in order to understand the step from the second premise to the conclusion. This, we should not diagram this argument by linking the two premises and drawing a single arrow to the conclusion, as in the examples considered so far. Rather each premise should have its own arrow pointing toward the conclusion.

Example 1.6. Diagram the following argument.

(1)[The Bensons must be home.] (2)[Their front door is open,] (3)[their car is in the driveway,] and (4)[their television is on,] **since** (5)[I can see its glow through the window.]

The argument is convergent. Statements 2, 3, and 4 function as independent reasons for the conclusion, statement 1. Each supports statement 1 separately, and must therefore be linked to it by a separate arrow.

Implicit Statements

Some arguments are incompletely expressed and can be thought of as having unstated assumptions. There are also cases in which it is clear that the author wishes the audience to draw an unstated conclusion. For example:

One of us must do the dishes, and it's not going to be me.

Here the speaker is clearly suggesting that the hearer should do the dishes, since no other possibility is left open.

Implicit premises or conclusions should be "read into" an argument only if they are required to complete the arguer's thought. The primary constraint governing interpolation of premises and conclusions is the _principle of charity_: in formulating implicit statements, give the arguer the benefit of the doubt and try to make the argument as strong as possible while remaining faithful to what you know of the arguer's thought. The goal is to minimize misinterpretation, whether deliberate or accidental.

Example 1.7. Complete and diagram the following incomplete argument.

(1)[If you were my friend, you wouldn't talk behind my back.]

This sentence suggests both an unstated premise and an unstated conclusion. The premise is:
(2)[You do talk behind my back.]
And the conclusion is:
(3)[You aren't my friend.]
Thus the diagram is:

$$\frac{1 + 2}{\downarrow}$$
$$3$$

Use and Mention

In any subject matter that deals extensively with language, confusion can arise as to whether an expression is being used to say something or mentioned as the subject matter of what is being said. To prevent this confusion, when an expression is mentioned rather than used, it should be enclosed in quotation marks. The following sentences both correctly employ this convention and are both true:

Socrates was a Greek philosopher.
"Socrates" is a name containing eight letters.

In contrast, the following sentences are false because of their incorrect use of quotation marks:

"Socrates" was a Greek philosopher.
Socrates is a name containing eight letters.

Example 1.8. Supply quotation marks in the following sentences in such a way as to make them true:

(a) Bill's name is Bill.
(b) The fact that $x+y = y+x$ is expressed by the equation $x+y = y+x$.

(a) Bill's name is "Bill".
(b) The fact that $x+y = y+x$ is expressed by the equation "$x+y = y+x$".

Formal vs. Informal Logic

Logic may be studied from two points of view. *Formal logic* is the study of *argument forms*, abstract patterns common to many different arguments. An argument form is something more than just the structure exhibited by an argument diagram, for it encodes something about the internal composition of the premises and conclusion. For example, the following is an argument form with "P" and "Q" denoting arbitrary statements.

If P, then Q
P
$\therefore Q$

In contrast, *informal logic* is the study of particular arguments in natural languages and the contexts in which they occur. Whereas formal logic emphasizes generality and theory, informal logic concentrates on practical argument analysis.

The two approaches are not opposed, but rather complement one another. In this book, the approach of Chapters 1, 2, 8, and 9 is predominantly informal. Chapters 3, 4, 5, 6, and 7 exemplify a predominantly formal point of view.

Chapter 2
ARGUMENT EVALUATION

IN THIS CHAPTER:

✔ *Evaluative Criteria*
✔ *Truth of Premises*
✔ *Validity and Inductive Probability*
✔ *Relevance*
✔ *The Requirement of Total Evidence*

Evaluative Criteria

Although an argument may have many objectives, its chief purpose is usually to demonstrate that a conclusion is true or at least likely to be true. Typically, then, arguments may be judged better or worse to the extent that they accomplish or fail to accomplish this purpose. In this chapter we examine four criteria for making such judgments:

Criterion 1. Whether all the premises are true

Criterion 2. Whether the conclusion is at least probable, given the truth of the premises

Criterion 3. Whether the premises are relevant to the conclusion

Criterion 4. Whether the conclusion is vulnerable to new evidence.

11

Truth of Premises

Criterion 1 is not by itself adequate for argument evaluation, but it provides a good start: no matter how good an argument is, it cannot establish the truth of its conclusion if any of its premises are false.

Example 2.1. Evaluate the following argument with respect to Criterion 1:

> Since all Americans today are isolationists, history will record that at the beginning of the twenty-first century the United States failed as a defender of world democracy.

> The premise "all Americans today are isolationists" is false, and so the argument does not establish the conclusion. Note that this does not mean the conclusion is false, only that the argument is of no use in determining the conclusion's truth or falsity.

In practice, an argument successfully communicates the truth of its conclusion only if those to whom it is addressed *know* that its premises are true. If the truth or falsity of one or more premises is unknown, the argument fails to establish its conclusion *so far as we know*. When we lack sufficient information to apply Criterion 1 reliably, we should suspend judgment until further information is acquired.

Example 2.2. A window has been broken. A little girl offers the following argument (expressed here in standard form):

> I saw Billy break the window.
> ∴ Billy broke the window.

Suppose we have reason to suspect the child did not see this. Evaluate the argument with respect to Criterion 1.

> Even if the child is telling the truth, her argument fails to establish its conclusion so long as we do not know that its premise is true. The best choice is to suspend judgment and seek more evidence.

Another limitation of Criterion 1 is that the truth of the premises— or that they are known to be true—provides no guarantee that the conclusion is also true. This is a necessary condition for establishing the conclusion, but, in a good argument, the premises must also *support* the conclusion.

Example 2.3. Evaluate the following argument with respect to Criterion 1:

> All acts of murder are acts of killing.
> ∴ Soldiers who kill in battle are murderers.

Since the premise is true, the argument satisfies Criterion 1. However, the premise fails to establish its conclusion, because the premise leaves open the possibility that some kinds of killing are not murder. This premise, though true, does not adequately support the conclusion, and so the argument proves nothing.

These examples demonstrate the need for further criteria for argument evaluation, in order to assess the degree to which a set of premises provides direct evidence for the conclusion.

Validity and Inductive Probability

Criterion 2 evaluates arguments with respect to the probability of the conclusion given the truth of the premises. In this respect, arguments may be classified into two categories. In a *deductive argument*, it is impossible for the conclusion of the argument to be false while its premises are all true. In an *inductive argument*, there is a certain probability that the conclusion is true if the premises are true, but there is also a probability that the conclusion is false.

Traditionally, the term "deductive" includes any argument that is intended to be deductive. *Valid deductive arguments* are those for which the conclusion cannot be false so long as the premises are true. *Invalid deductive arguments* are arguments that purport to be deductive but can have a false conclusion even with true premises.

Example 2.4. Classify the following arguments as either deductive or inductive.

(a) No mortal can halt the passage of time.
     You are mortal.
     ∴ You cannot halt the passage of time.
(b) It is usually cloudy when it rains.
     It is raining now.
     ∴ It is cloudy now.
(c) Some pigs have wings.
     All winged things sing.
     ∴ Some pigs sing.

    (a) Deductive
    (b) Inductive
    (c) Deductive

 **Note!**

When we say that it is impossible for the conclusion of a deductive argument to be false while its premises are true, the term "impossible" is to be understood in the very strong sense of *logically impossible.*

A deductive argument whose premises are all true is said to be *sound* and establishes with certainty that its conclusion is true. For example, the argument in Example 2.4(a) is sound. Although deductive arguments provide the greatest certainty, in practice we must often settle for inductive arguments, which have a range of inductive probabilities and hence vary widely in reliability.

Example 2.5. Evaluate the reasoning of the following arguments:

(a) Visitors to China almost never contract malaria there.
     Jan is visiting China.
     ∴ Jan will not contract malaria there.
(b) I dream about monsters.
     My brother dreams about monsters.
     ∴ Everyone dreams about monsters.

The phrase "almost never" in the first premise of argument (a) lends a strong inductive probability to this argument. In contrast, argument (b) has a weak inductive probability since it generalizes from a small sample of two individuals to the largest possible population of everyone.

There is no sharp line between strong and weak inductive reasoning, since what is strong for one purpose may not be strong enough for another. For example, if the conclusion that a certain valve will not malfunction over a 5-year period has a 0.9 probability, we may regard this reasoning as strong. But if the valve is part of a nuclear power plant and the lives of thousands of people depend on its functioning correctly, then 0.9 may not be strong enough. Thus, there is no simple answer to the question: "How high an inductive probability must an argument have for the reasoning to be classified as strong?"

The relation of the inductive probability of a complex argument to the inductive probabilities of its component steps is in general a very intricate affair. However, there are a few helpful rules of thumb:

1. With regard to complex nonconvergent arguments, if one or more of the steps is weak, then usually the inductive probability of the argument as a whole is low.
2. If all the steps of a complex nonconvergent argument are strongly inductive or deductive, then the inductive probability of the whole is usually fairly high.
3. The inductive probability of a convergent argument is usually as high as the inductive probability of its strongest branch.
4. If all the steps of a complex argument are deductive, then so is the argument as a whole.

Relevance

Not every argument with true premises and high inductive probability is a good argument. Any argument that lacks relevance does not prove the truth of its conclusion and is said to commit the *fallacy of relevance*. This is our third evaluative criterion for arguments, and as with inductive probability, relevance is a matter of degree.

Example 2.6. Evaluate the following argument with respect to inductive probability and degree of relevance.

> I abhor the idea of an infinitely powerful creator.
> ∴God does not exist.

> Personal likes or dislikes have nothing to do with the actual existence of God; hence the premise is not relevant. The inductive probability of this argument is also not very high.

Relevance and inductive probability do not always vary together. The simplest cases of high inductive probability with low relevance occur among arguments whose conclusions are logically necessary. A *logically necessary* statement is a statement whose very conception or meaning requires its truth. Consider the following examples:

> Either something exists, or nothing at all exists.
> $2 + 2 = 4$.

You Should Know

If a logically necessary statement is the conclusion of an argument, then the argument is automatically deductive and no premise is needed to believe the conclusion, regardless of the nature of the premises.

Example 2.7. Evaluate the following argument with respect to inductive probability and degree of relevance.

> Some sheep are black.
> Some sheep are white.
> ∴If something is a cat, then it is a cat.

> This artificial argument has a logically necessary conclusion and is deductive, even though its premises are irrelevant to its conclusion and provide no reason to believe the conclusion.

Example 2.8. Evaluate the following argument with respect to inductive probability and degree of relevance.

>All of Fred's friends go to Freeport High.
>All of Freda's friends go to Furman High.
>Nobody goes to both Freeport and Furman.
>∴ Fred and Freda have no friends in common.

The argument is deductive, and so has the maximum inductive probability. The premises are highly relevant to its conclusion.

An argument can be deductive and yet lack relevance if it has inconsistent premises. A set of statements is *inconsistent* if it is logically impossible for all of them to be true simultaneously. For example, the following set of statements is inconsistent:

>Jim is taller than Bob.
>Bob is taller than Sally.
>Sally is taller than Jim.

Any argument with inconsistent premises is deductive. Since inconsistent premises cannot all be true, it is impossible for the premises to be true while some conclusion is false. Hence, *any conclusion follows deductively from inconsistent premises*. However, no such argument is sound and *no conclusion is ever proved by deducing it from inconsistent premises.*

Example 2.9. Evaluate the following argument with respect to inductive probability and degree of relevance.

>This book has more than 900 pages.
>This book has fewer than 800 pages.
>∴ This is a very profound book.

Since it is logically impossible for the book to have more than 900 pages and fewer than 800 pages, it is impossible for both premises to be true while the conclusion is false. Therefore the argument is deductive and has the maximum inductive probability. However, the premises are irrelevant to the conclusion.

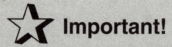 **Important!**

A good argument requires true premises (Criterion 1), high inductive probability (Criterion 2), and a high degree of relevance (Criterion 3).

The Requirement of Total Evidence

Inductive arguments differ from deductive arguments in the crucial aspect of their vulnerability to new evidence. A deductive argument remains deductive if new premises are added (regardless of the nature of the premises). In contrast, an inductive argument can be either strengthened or weakened by the addition of new premises.

Example 2.10. Evaluate the following argument with respect to inductive probability and degree of relevance.

> Very few Russians speak English well.
> Sergei is Russian.
> ∴ Sergei does not speak English well.

This argument has high inductive probability, and its premises are relevant to its conclusion.

Example 2.11. Evaluate the following argument with respect to inductive probability and degree of relevance.

> Very few Russians speak English well.
> Sergei is Russian.
> Sergei is an exchange student at an American university.
> Exchange students at American universities almost always
>   speak English well.
> ∴ Sergei does not speak English well.

The premises are relevant to the conclusion. However, the argument has low inductive probability since the first two premises support the conclusion, while the last two support its negation.

Example 2.12. Evaluate the following argument with respect to inductive probability and degree of relevance.

Very few Russians speak English well.
Sergei is Russian.
Sergei is an exchange student at an American university.
Exchange students at American universities almost always
    speak English well.
Sergei is a mute.
∴ Sergei does not speak English well.

The argument is now deductive, since the conclusion follows from the fifth premise, and so has the maximum inductive probability. The premises are all relevant to the conclusion, although once we have the fifth premise, the first four premises are superfluous.

Additional premises cannot alter the maximum inductive probability of a deductive argument. If all of the premises are true, then the conclusion is true and no further evidence can decrease its certainty. In contrast, inductive arguments can be converted into arguments with higher or lower inductive probability by the addition of premises, and the choice of premises is crucial in inductive reasoning. In fact, by selectively manipulating the evidence, we may be able to make a conclusion appear as probable or as improbable as we would like.

The selective manipulation of the evidence is illegitimate and defines our fourth criterion of argument evaluation: the *requirement of total evidence*. This criterion asserts that if an argument is inductive, its premises must contain all known evidence relevant to the conclusion. Inductive arguments that fail to meet this requirement commit the *fallacy of suppressed evidence*.

Example 2.1. Evaluate the following argument with respect to the four criteria.

Most cats do well in apartments.
They are very affectionate and love being petted.
∴ This cat will make a good pet.

The argument fares well with the first three criteria: the premises are true and relevant to the conclusion and a high inductive probability.

However, if the arguer is withholding the fact that the cat grew up in a cat shelter, where it became dirty and aggressive, the argument is flawed by the fallacy of suppressed evidence.

Remember

Even if an inductive argument meets the requirement of total evidence, it may still lead us from true premises to a false conclusion. *Inductive arguments provide no guarantees.*

Suppressed evidence should not be confused with implicit premises. Implicit premises are assumptions that the author of an argument intends the audience to take for granted. In contrast, suppressed evidence is information that the author has deliberately concealed or intentionally omitted. Implicit assumptions are part of the author's argument, while suppressed evidence is not.

Some logicians argue that the requirement of total evidence is too stringent. Indeed, arguments on complex issues will generally suffer to some degree from the fallacy of suppressed evidence. The best arguments are those minimizing suppressed evidence and suppressing no evidence that drastically affects the probability of the conclusion.

Chapter 3
PROPOSITIONAL LOGIC

IN THIS CHAPTER:

✔ *Argument Forms*
✔ *Logical Operators*
✔ *Formalization*
✔ *Semantics of the Logical Operators*
✔ *Truth Tables for Wffs*
✔ *Truth Tables for Argument Forms*
✔ *Refutation Trees*

Argument Forms

Formal logic is the study of *argument forms*, abstract patterns of reasoning shared by many different arguments. Our motivation is the idea that a valid deductive argument is one whose conclusion cannot be false while its premises are all true. By studying argument forms, we shall be able to precisely characterize this idea. For example, the following two arguments have the same form:

1. Today is either Monday or Tuesday.
   Today is not Monday.
   ∴ Today is Tuesday

21

2. Either Rembrandt painted the *Mona Lisa* or Michelangelo did.
   Rembrandt didn't do it.
   ∴ Michelangelo did

Both of these arguments are deductively valid. Their common form is known by logicians as *disjunctive syllogism*, and can be represented as:

Either *P* or *Q*
It is not the case that *P*
∴ *Q*

In this representation, the letters *P* and *Q* are called *sentence letters* and function as placeholders for declarative sentences. Every argument with this form may be obtained by replacing these sentence letters with statements, where each occurrence of the same letter is replaced by the same statement. For example, argument 1 is an *instance* of disjunctive syllogism obtained from the argument form above by replacing *P* with "Today is Monday" and *Q* with "Today is Tuesday".

Example 3.1. Identify the argument form of the following two arguments.

(a) If today is Monday, then I have to go to the dentist.
    Today is Monday
    ∴ I have to go to the dentist.
(b) If you and Jane passed the test, then so did Olaf.
    Olaf did not pass.
    ∴ It's false that both you and Jane passed the test.

Argument (a) has the argument form known as *modus ponens*:
If *P*, then *Q*
P
∴ *Q*

Argument (b) has the argument form known as *modus tollens*:
If *P*, then *Q*
It's not the case that *Q*
∴ It's not the case that *P*

If every instance of an argument form is valid (cannot have true premises and a false conclusion), the argument form itself is said to be *valid*; otherwise the argument form is said to be *invalid*. For example, disjunctive syllogism, modus ponens, and modus tollens are valid argument forms. In contrast, the following argument form (known as *affirming the consequent*) is invalid:

> If *P*, then *Q*
> *Q*
> ∴ *P*

The following instance of this argument form is invalid and provides a *counterexample* to the validity of affirming the consequent.

> If you are dancing on the moon, then you are alive.
> You are alive.
> ∴ You are dancing on the moon.

Remember

An argument form with even one invalid instance is invalid.

Logical Operators

We shall be concerned with those argument forms consisting of sentence letters combined with one or more of the expressions: "it is not the case that," "and," "either . . . or," "if . . . then," and "if and only if." These expressions are called *logical operators* or *connectives* and many different forms may be constructed from these simple expressions.

The operator "it is not the case that" prefixes a statement to form a new statement called the *negation* of the first. For example, "It is not the case that he is a smoker" is the negation of "He is a smoker". In English, there are many variations of negation; "He is a nonsmoker" and "He is not a smoker" are also negations of "He is a smoker".

The other four logical operators are *binary* operators joining two statements into a compound statement. A *conjunction* is two statements joined by "and." Conjunction is also expressed by "both . . . and," "but," "yet," "although," "nevertheless," "whereas," and "moreover," all of which affirm both statements they join.

A *disjunction* consists of two statements joined by "either . . . or", although the term "either" is often omitted. For example, "Today is either Monday or Tuesday" and "Today is Monday or Tuesday" express the same idea.

Conditionals are statements formed using "if . . . then." The statement following "if" is called the *antecedent* and the other statement is the *consequent*. In the conditional "If you touch me, then I'll scream", "You touch me" is the antecedent and "I'll scream" is the consequent. In conditionals, the word "then" may be omitted and they can also be stated in reverse order; e.g., "I'll scream if you touch me".

Finally, *biconditionals* are statements formed using "if and only if". Biconditionals may be regarded as conjunctions of two conditionals (thus the name). For example, the statement "It is a triangle if and only if it is a three-sided polygon" is a variant of the statement "If it is a three-sided polygon, then it is a triangle; and if it is a triangle, then it is a three-sided polygon".

Example 3.2. Classify the following sentences by logical operator.

(a) If today is Monday, yesterday was Sunday.
(b) Today is Monday if yesterday was Sunday.
(c) It's her birthday if and only if it's Monday.
(d) It is either Monday or Sunday.
(e) John and Mary are competent.
(f) He is incompetent.

(a) conditional with antecedent: "Today is Monday"
(b) conditional with antecedent: "Yesterday was Sunday"
(c) biconditional
(d) disjunction
(e) conjunction of "John is competent" and "Mary is competent"
(f) negation of "He is competent"

To facilitate recognition and comparison of argument forms each logical operator is represented by a special symbol:

Logical Operator	Symbol
it is not the case that	~
and	& •
either . . . or	∨
if . . . then	→ ⊃
if and only if	↔ ≡

For example, using these symbols, the argument form of disjunctive syllogism may be expressed as:

$P \vee Q$
$\sim P$
$\therefore Q$

We also write argument forms horizontally, with the premises separated by commas and the conclusion denoted by the *turnstile* "⊢". Following these conventions, disjunctive syllogism is expressed as:

$P \vee Q, \sim P \vdash Q$

Example 3.3. Express the argument form modus tollens in symbolic notation.

If P, then Q
It is not the case that Q
$\therefore$ It is not the case that P

$P \rightarrow Q, \sim Q \vdash \sim P$

Formalization

The language consisting of the symbolic notation of the previous section is called the *language of propositional logic*. We examine the syntax of this language by showing how various English sentences may be expressed, or *formalized*, as symbolic formulas, and then state the explicit grammatical rules, or *formation rules*, for the language.

The process of formalization translates an English sentence or argument into a sentence form or argument form, a structure composed of sen-

tence letters and logical operators. Whenever we talk
about the meaning of a sentence letter, we are speak-
ing of its meaning *under the particular interpretation
specified by the problem at hand*. For example, if we
interpret the sentence letter M as "Today is Monday",
then ~ M formally expresses the sentence "Today is
not Monday".

Example 3.4. Interpreting the sentence letter R as "It is raining" and the
letter S as "It is snowing", formalize each sentence in the language of
propositional logic.

    (a) It is either raining or snowing.
    (b) It is raining, but it is not snowing.
    (c) If it is not raining, then it is snowing.
    (d) It is not the case that if it is raining, then it is snowing.
    (e) It is neither raining nor snowing.
    (f) Either it is raining, or it's both snowing and raining.

    (a) $R \vee S$
    (b) $R \,\&\, (\sim S)$
    (c) $(\sim R) \rightarrow S$
    (d) $\sim (R \rightarrow S)$
    (e) both $\sim (R \vee S)$ and $(\sim R) \,\&\, (\sim S)$ are correct
    (f) $R \vee (S \,\&\, R)$

Formulas are constructed from the following three sets of symbols
which constitute the *vocabulary* of the language of propositional logic:

    Sentence letters: Every capital letter is a sentence letter; we may
    also add subscripts, for example, S_1, S_2, S_3, etc.
    Logical Operators: ~, &, $\vee$, $\rightarrow$, $\leftrightarrow$
    Parentheses: (,)

The *logical* symbols of our formal language are the logical operators and
the parentheses. We call sentence letters the *nonlogical* symbols.
    A *formula* of the language of propositional logic is any sequence of
symbols of the vocabulary. The answers in Example 3.4 are all formulas,
but so are such nonsense sequences as ((&(P. We distinguish nonsense

sequences from meaningful formulas (known as *well-formed formulas* or *wffs*) via the following *formation rules*. These rules use Greek letters to denote arbitrary formulas.

Rule 1. Any sentence letter is a wff.
Rule 2. If φ is a wff, then $\sim \varphi$ is a wff.
Rule 3. If φ and ψ are wffs, then $(\varphi \ \& \ \psi)$, $(\varphi \vee \psi)$, $(\varphi \to \psi)$, and $(\varphi \leftrightarrow \psi)$ are wffs.
Rule 4. Anything not asserted to be a wff by Rules 1 through 3 is not a wff.

Complex wffs are built up from simple ones by repeated applications of the formation rules. For example, by Rule 1, P and Q are both wffs, by Rule 3, *(P & Q)* is a wff, and so, by Rule 2, $\sim$ *(P & Q)* is a wff. A *subwff* is a part of a wff which is itself a wff. For example, P is a subwff of $\sim$ *(P & Q)*, and $\sim R$ is a subwff of $\sim \sim R$.

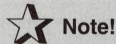 **Note!**

We adopt the unofficial convention that outer parentheses may be omitted. For example, we write $R \vee S$ rather than $(R \vee S)$.

Example 3.5. Use the formation rules to determine if the following formulas are wffs or not. Explain your answer.

(a) *PQ*
(b) $P \to Q$
(c) *((P & Q) → R)*
(d) *(P ∨ Q ∨ R)*

(a) Not a wff. By Rule 3, two or more sentence letters produce a wff only in combination with a binary operator.
(b) Not officially a wff, since outer brackets are missing. But, this is unofficially a wff according to our convention.
(c) P, Q, R are wffs by Rule 1, *(P & Q)* is a wff by Rule 3, and so *((P & Q) → R)* is a wff by Rule 3.
(d) Not a wff. We can only combine two sentence letters at a time.

Example 3.6. Formalize the following argument:

> If today is Saturday, then tomorrow is Sunday. If tomorrow is Sunday, then the day after tomorrow is Monday. Therefore, if today is Saturday, then the day after tomorrow is Monday.

$$S \rightarrow U, U \rightarrow M \vdash S \rightarrow M$$

Semantics of the Logical Operators

The semantics of an expression is its contribution to the truth or falsity, that is, the *truth value*, of the statements in which it occurs. The semantics of a logical operator is given by a rule for determining the truth value of any compound sentence involving that operator, based on the truth values of the components. We assume the *principle of bivalence*: that *true* and *false* are the only truth values and that in every possible situation each statement has one and  only one of them. We use the abbreviations "T" for "true" and "F" for "false."

For the semantic rule for negation, we observe that the negation of a sentence φ is true if φ is false, and φ is false if φ is true. We use our abbreviations to summarize this rule in the following truth table:

φ	$\sim \varphi$
T	F
F	T

We list under φ the two possible truth values T and F. Given bivalence, these are the only two possibilities, and the table completely describes the truth value of $\sim\varphi$ in every situation.

A conjunction is true if both of its conjuncts are true, and false otherwise. Since conjunction operates on two statements, there are four possible situations to consider as represented by the four horizontal rows of the following truth table for conjunction:

φ	ψ	$\varphi \mathbin{\&} \psi$
T	T	T
T	F	F
F	T	F
F	F	F

A disjunction is true if one or both of its disjuncts is true, and false only if both of disjuncts are false:

φ	ψ	$\varphi \vee \psi$
T	T	T
T	F	T
F	T	T
F	F	F

There is an *exclusive* sense of disjunction in which "either P or Q" means "either P or Q, but *not* both". In English, this exclusive sense is quite common, especially when "or" is preceded by "either." For example, if the boss says "You may take either Monday or Wednesday off", what's meant is undoubtedly the exclusive disjunction and the best formalization is $(M \vee W) \mathbin{\&} \sim (M \mathbin{\&} W)$. As is standard in logic, we adopt the inclusive sense of disjunction symbolized by $\vee$ and characterized by the truth table above. For example, the sentence "She must be intelligent or rich" does not exclude the possibility that the person is *both* intelligent and rich and is best formalized as $I \vee R$.

The conditional statement $P \rightarrow Q$ asserts that: "it is not the case that P and not Q". This statement has the form $\sim (P \mathbin{\&} \sim Q)$ and we obtain the truth table for $P \rightarrow Q$ by finding the truth table for the wff $\sim (P \mathbin{\&} \sim Q)$.

You Need to Know

To construct a truth table for a complex wff, we find the truth values for its smallest subwffs and then use the truth tables for the logical operators to calculate values for increasingly larger subwffs until we obtain the values of the whole wff.

For the conditional, we construct the following truth table:

P	Q	$\sim Q$	$P \ \& \sim Q$	$\sim (P \ \& \sim Q)$
T	T	F	F	T
T	F	T	T	F
F	T	F	F	T
F	F	T	F	T

Therefore, the conditional is false if its antecedent is true and its consequent is false, and otherwise the conditional is true, as summarized in the following truth table:

φ	ψ	$\varphi \to \psi$
T	T	T
T	F	F
F	T	T
F	F	T

Finally, we consider the biconditional $P \leftrightarrow Q$, which means the same thing as $(P \to Q) \ \& \ (Q \to P)$. As above, we can construct the truth table of this complex statement using the basic truth tables. When we do so, we find that a biconditional is true if its two components have the same truth value and false if their truth values differ, as summarized in the following truth table:

φ	ψ	$\varphi \leftrightarrow \psi$
T	T	T
T	F	F
F	T	F
F	F	T

As suggested by our definition of the semantic rules for conditionals and biconditionals, these logical operators are unnecessary in the sense that every statement of the form $P \to Q$ and $P \leftrightarrow Q$ can be expressed in terms of $\sim$, $\&$, and $\vee$. In fact, negation together with any one of conjunction, disjunction, or conditional, can express every other logical operator represented by a truth table; our set of logical operators is said to be *functionally complete*.

Truth Tables for Wffs

Our presentation of the semantic rules for $\rightarrow$ and $\leftrightarrow$ provided initial examples for the construction of truth tables for complex wffs. In this section we discuss this procedure more systematically.

To create a truth table for a wff, we write the sentence letters appearing in the wff in alphabetical order as the left columns of the truth table. If there are n sentence letters, there are 2^n rows in the truth table representing all possible distinct assignments of T and F to the letters as illustrated in the following examples. Using the truth tables for the logical operators, we calculate the truth values of the wff by determining the value of its smallest subwffs first and then increasingly larger subwffs, until we obtain the values for the whole wff.

Example 3.7. Construct the truth table for the wff $\sim \sim P$.

Since $\sim \sim P$ contains one sentence letter, its truth table has $2^1 = 2$ rows. We apply the negation table twice, switching the truth values twice and obtaining the final truth values in column 3.

P	$\sim$	$\sim$	P
T	T	F	T
F	F	T	F
	3	2	1

Example 3.8. Construct the truth table for the wff $(\sim P) \vee Q$.

Since $(\sim P) \vee Q$ contains two sentence letters, its truth table has $2^2 = 4$ rows. We first copy the Q column under Q (column 1) and apply the negation table to the P column to obtain $\sim P$ (column 2). We then determine the truth values for $(\sim P) \vee Q$ in column 3 from those for $\sim P$ and Q, using the basic disjunction table. A disjunction is false $\vee$ if and only if both disjuncts are false, which occurs only at line 2. Thus, $(\sim P) \vee Q$ is false at line 2 and true at all other lines. The final truth table for $(\sim P) \vee Q$ is:

P	Q	$\sim P$	$\vee$	Q
T	T	F	T	T
T	F	F	F	F
F	T	T	T	T
F	F	T	T	F
		2	3	1

Example 3.9. Construct the truth tables for the wff $P \vee \sim P$ and the wff $P \ \& \sim P$.

Both wffs contain one sentence letter, and so both truth tables have 2 rows. We first copy P (column 1) and then apply the negation table to obtain $\sim P$ (column 2). Finally, we use the basic disjunction and conjunction tables on columns 1 and 2 to obtain the final truth values in column 3.

P	P	$\vee$	$\sim P$
T	T	T	F
F	F	T	T
	1	3	2

P	P	$\&$	$\sim P$
T	T	F	F
F	F	F	T
	1	3	2

Wffs which, like $P \vee \sim P$, are true at every line in their truth table are called *tautologies*, as are all specific statements of the same form. Similarly wffs which, like $P \ \& \sim P$, are false at every line of their truth table are called *contradictions* and are said to be *truth-functionally inconsistent*, as are all specific statements of the same form.

Example 3.10. Construct the truth table for the wff $P \rightarrow (Q \vee \sim R)$.

Since $P \rightarrow (Q \vee \sim R)$ contains three sentence letters, its truth table has $2^3 = 8$ rows. We begin as before, and in the end apply the basic implication table to column 1 and column 4 to obtain the final truth values for $P \rightarrow (Q \vee \sim R)$ in column 5.

P	Q	R	P	→	(Q	∨	~R)
T	T	T	T	T	T	T	F
T	T	F	T	T	T	T	T
T	F	T	T	F	F	F	F
T	F	F	T	T	F	T	T
F	T	T	F	T	T	T	F
F	T	F	F	T	T	T	T
F	F	T	F	T	F	F	F
F	F	F	F	T	F	T	T
			1	5	2	4	3

Formulas which, like $P \to (Q \vee \sim R)$, are true at some lines of their truth tables and false at others are said to be *truth-functionally contingent*, as are all specific statements of the same form. A contingent statement is one that could be either true or false, *so far as the operators of propositional logic are concerned*. However, some truth-functionally contingent statements are not actually contingent. For example the statement "Jim is a bachelor and (the same) Jim is married" has propositional form $B \ \& \ M$ and is truth-functionally contingent. However, this sentence is not contingent, but inconsistent, as a result of the semantics of the expressions "is a bachelor" and "is married," in addition to the logical operator "and."

Truth Tables for Argument Forms

We provide a rigorous account of deductive validity based on the semantics of the logical operators. Recall that an argument form is valid if and only if all its instances are valid and that an instance of a form is valid if it is impossible for its conclusion to be false while its premises are true. If we put not just a single wff on a truth table, but an entire argument form, we can use the table to determine its validity.

Example 3.11. Demonstrate the validity of the disjunctive syllogism:

Either the princess or the queen attends the ceremony.
The princess does not attend.
∴ The queen attends.

We formalize this argument as $P \vee Q, \sim P \vdash Q$ and construct the corresponding truth table for this argument:

P	Q	P	∨	Q,	~P	⊢	Q
T	T	T	T	T	F		T
T	F	T	T	F	F		F
F	T	F	T	T	T		T
F	F	F	F	F	T		F

This table is computed in the same way as the tables for a single wff, except that it displays three wffs, instead of just one. Note that only in the third line of the truth table are both premises true and that, in this line, the conclusion is true as well. Thus, it is not possible for all the premises to be true and the conclusion false, and so the table shows that this argument, and every argument of this form, is valid.

If an argument form is *invalid*, the corresponding truth table shows the invalidity by exhibiting one or more lines (a *counterexample*) in which all the premises have value T while the conclusion has value F. Counterexamples provide a guide for constructing instances of the argument form with true premises and a false conclusion; the existence of even one counterexample is sufficient to establish invalidity.

Example 3.12. Demonstrate the invalidity of the converse error (also known as affirming the consequent):

$P \rightarrow Q, Q \vdash P.$

We first construct the following truth table for this argument form:

P	Q	P	→	Q,	Q	⊢	P	
T	T	T	T	T	T		T	
T	F	T	F	F	F		T	
F	T	F	T	T	T		F	×
F	F	F	T	F	F		F	

The table shows the two possible situations in which the premises are both true in the first and third lines. At the first line, the conclusion is also true; but at the third line, it is false. Thus, the third line is a counterexample (the "×" to the right indicates this).

Remember

We determine the validity of an argument form of propositional logic by putting the entire form on a truth table. If the table displays no counterexample, the form is valid, and if the table displays one or more counterexamples, the form is invalid.

Refutation Trees

Truth tables provide a rigorous and complete test for the validity or invalidity of propositional logic argument forms. Indeed, they provide an *algorithm*, a precisely specifiable test that can be performed by a computer and always yields an answer after a finite number of finite operations. We say a formal system is *decidable* when there is such an algorithm for determining the validity of argument forms expressible in the system. Thus, truth tables ensure the decidability of propositional logic. But they are also cumbersome and inefficient, especially for problems involving three or more sentence letters. Refutation trees provide a more efficient algorithm for performing the same task.

A refutation tree is an exhaustive search for the ways in which all wffs in a given list can be true. To test an argument form for validity using a refutation tree, we construct a list consisting of its premises and the *negation* of its conclusion. The search is carried out by breaking down the wffs on the list into sentence letters or their negations. If we can find an assignment of truth and falsity to sentence letters that makes all the wffs on the list true, then under that assignment the premises of the form are true while its conclusion is false and the argument form is invalid. If the search turns up no assignment of truth and falsity to sentence letters that makes all the wffs on the list true, then our attempted refutation has failed and the form is valid.

Example 3.13. Construct a refutation tree to demonstrate the validity of the argument form: $P \& Q \vdash \sim \sim P$.

We list the premise and the negation of the conclusion:

$P \& Q$
$\sim \sim \sim P$

The premise is true if and only if P and Q are both true. Therefore, we replace $P \& Q$ by writing P and Q at the bottom of the list and checking off $P \& Q$ to obtain:

✓ $P \& Q$
         $\sim \sim \sim P$
         P
         Q

Since $\sim \sim \sim P$ is true if and only if the simpler wff $\sim P$ is true, we check off $\sim \sim \sim P$ and replace it with $\sim P$ to obtain:

✓ $P \& Q$
✓ $\sim \sim \sim P$
         P
         Q
         $\sim P$
         X

Every wff has been broken down into sentence letters or their negations, all of which must be true for every member of our original list to be true. However, P and $\sim P$ both appear on the final list and they cannot simultaneously be true. Since it is not possible for every wff in the list to be true, we write an X at the bottom. Our search for a refutation of the original argument has failed and this argument form is valid.

Example 3.14. Construct a refutation tree to demonstrate the invalidity of the argument form: $P \vee Q, P \vdash \sim Q$.

We list the two premises and the negation of the conclusion:

$P \lor Q$
P
$\sim \sim Q$

The wff $\sim \sim Q$ is equivalent to the simpler wff Q, so we check off $\sim \sim Q$ and write Q at the bottom of the list.

$P \lor \text{Q}$
P
✓ $\sim \sim Q$
Q

The wff $P \lor Q$ is true if and only if either of P or Q is true, so we check off $P \lor Q$ and branch the tree as follows:

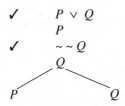

✓ $P \lor Q$
P
✓ $\sim \sim Q$
Q

P Q

Since every wff has been broken down into sentence letters or their negations, the refutation tree is complete. Neither path through the tree contains both a sentence letter and its negation (we did not close either path with an X), so both paths represent possible truth value assignments for which all three wffs in the original list of premises and negated conclusion are simultaneously true. In fact, P and Q appear on each path. If we take both to be true, we obtain a specific truth value assignment for which the premises of the argument form $P \lor Q, P \vdash \sim Q$ are all true, but the conclusion is false. This counterexample shows the form is invalid.

In a refutation tree a list of wffs is broken down (based on the logical operators) into sentence letters or their negations representing the ways in which the wffs on the original list may be true. Wffs with different logical operators have different rules for extending refutation trees; the following is a comprehensive list of rules for extending trees along with their standard abbreviations.

You Need to Know

To test an argument form for validity using a refutation tree, we list the premises and the negation of the conclusion. If every path of the corresponding finished tree is closed (e.g., Example 3.13), the form is valid; if one or more paths of the finished tree is open (e.g., Example 3.14), the form is invalid.

Negation (~): If an open path contains both a wff and its negation, write an X at the bottom of the path.

Negated Negation (~ ~): For an unchecked wff ~ ~φ, check it and write φ at the bottom of every open path containing it.

Conjunction (&): For an unchecked wff φ & ψ, check it and write both φ and ψ at the bottom of every open path containing it.

Negated Conjunction (~&): For an unchecked wff ~(φ & ψ), check it and branch the bottom of every open path containing it, with ~φ on one branch and ~ψ on the other.

Disjunction ($\vee$): For an unchecked wff $\varphi \vee \psi$, check it and branch the bottom of every open path containing it, with φ on one branch and ψ on the other.

Negated Disjunction (~$\vee$): For an unchecked wff ~($\varphi \vee \psi$), check it and write both ~φ and ~ψ at the bottom of every open path containing it.

Conditional ($\rightarrow$): For an unchecked wff $\varphi \rightarrow \psi$, check it and branch the bottom of every open path containing it, with ~φ on one branch and ψ on the other.

Negated Conditional (~$\rightarrow$): For an unchecked wff ~($\varphi \rightarrow \psi$), check it and write both φ and ~ψ at the bottom of every open path containing it.

Biconditional ($\leftrightarrow$): For an unchecked wff $\varphi \leftrightarrow \psi$, check it and branch the bottom of every open path containing it, with φ and ψ on one branch and ~φ and ~ψ on the other branch.

Negated Biconditional (~↔): For an unchecked wff ~(φ ↔ ψ), check it and branch the bottom of every open path containing it, with φ and ~ψ on one branch and ~φ and ψ on the other.

An *open* path in a refutation tree is a path that has not been ended with an X; paths ending with an X are said to be *closed*. A path is *finished* if it is closed or if the only unchecked wffs on the path are sentence letters or their negations (so none of the extension rules apply to wffs on the path), and a tree is *finished* if all of its paths are finished.

In the following examples, we annotate trees by numbering the lines and stating the rules and lines used to add wffs to the tree.

Example 3.15. Construct a refutation tree to determine the validity of the argument form:

$P{\to}Q, Q{\to}R, P \vdash R$.

We first list the premises and the negation of the conclusion and then apply our rules for extending refutation trees as indicated. Since the finished tree is closed, the attempted refutation has failed and the argument form is valid.

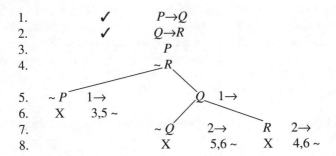

Example 3.16. Construct a refutation tree to determine the validity of the argument form:

$\sim P{\to}\sim Q \vdash Q{\to}P$.

We first list the premise and the negation of the conclusion and then apply our rules for extending refutation trees as indicated. The finished tree has two open paths, each indicating that the premise is true and the conclusion is false when P is true and Q is false. Therefore, this argument form is invalid.

1.   ✓ $\sim P \rightarrow \sim Q$
2.   ✓ $\sim (P \rightarrow Q)$
3. P $2 \sim \rightarrow$
4. $\sim Q$ $2 \sim \rightarrow$

5.   ✓ $\sim\sim P$ $1 \rightarrow$ $\sim Q$ $1 \rightarrow$
6. P $5 \sim\sim$

⭐ Note!

Every open path of the finished tree is a prescription for constructing a counterexample.

    Refutation trees can also be used to test for truth-functional consistency and tautologousness. A list of wffs is *truth-functionally consistent* if all of its members can be true simultaneously. If the corresponding refutation tree has at least one open path, then all the wffs on the list can be true simultaneously and the list is said to be consistent. If a finished tree contains no open paths, the list of formulas is inconsistent. For tautologousness, note that a wff is tautologous if and only if its negation is inconsistent. Therefore, any wff φ is tautologous if and only if all the paths on the finish tree for $\sim\varphi$ are closed.

Example 3.17. Construct a refutation tree to determine if the following wff is tautologous: $(P \rightarrow Q) \vee (P \,\&\, \sim Q)$.

    We list the negation of the wff, and then apply our rules for extending refutation trees as indicated. Since all the paths close, our attempt to find a way to make its negation true has failed and $(P \rightarrow Q) \vee (P \,\&\, \sim Q)$ is a tautology.

1. ✓ $\sim((P \rightarrow Q) \vee (P \,\&\, \sim Q))$
2. ✓ $\sim(P \rightarrow Q)$ $1 \sim \vee$
3. ✓ $\sim(P \,\&\, \sim Q)$ $1 \sim \vee$
4. P $2 \sim \rightarrow$
5. $\sim Q$ $2 \sim \rightarrow$

6. $\sim P$ $3 \sim \&$ $\sim \sim Q$ $3 \sim \&$
7. X $4, 6 \sim$ X $5, 6 \sim$

When constructing refutation trees, keep in mind the following:

1. The rules for constructing trees apply only to whole wffs, not to sub-wffs. For example, the following use of negated negation is not permissible:

   1. ✓ $P \,\&\, \sim \sim Q$
   2. $P \,\&\, Q$ $1 \sim\sim$ (incorrect)

2. The order in which rules are applied makes no difference to the final answer, but it is usually more efficient to apply non-branching rules first.
3. The open paths of a finished tree for an invalid argument form display all counterexamples to that argument form.

Chapter 4

THE PROPOSITIONAL CALCULUS

IN THIS CHAPTER:

- ✔ *The Notion of Inference*
- ✔ *Nonhypothetical Inference Rules*
- ✔ *Hypothetical Inference Rules*
- ✔ *Derived Rules*
- ✔ *Theorems*
- ✔ *Equivalences*

The Notion of Inference

Chapter 3 approached propositional logic from a semantic point of view, testing the validity of argument forms based on the intended interpretations of the logical operators. In this chapter, we establish deductive validity by *inferring* or *deriving* the conclusion from the premises; that is, we show that the conclusion *follows from* the premises. Typically, this is how one offers an argument to support a certain statement: one shows

how the conclusion can be reached through a finite number of successive steps of reasoning, each of which is fully explicit and indisputable.

The key to proving validity via step-by-step deductions lies in mastering the principles (called *rules of inference*) used to make each successive step. We shall formulate an introduction and an elimination rule of inference for each of the five logical operators. The *elimination rule* for an operator is used to reason from premises in which it is the main operator. The *introduction rule* for an operator is used to derive conclusions in which it is the main operator. These rules form a system for performing calculations with propositions called the *propositional calculus*. A *deduction* (or *derivation* or *proof*) in the propositional calculus is a series of wffs in the language of propositional logic, each of which is either a premise or the result of applying a rule of inference.

Nonhypothetical Inference Rules

To illustrate the concept of a derivation, consider the following argument form:

$$P \rightarrow Q, Q \rightarrow R, P \vdash R$$

Since this form is valid (see Example 3.15), the conclusion should be derivable from the premises; here is the derivation:

1.	$P \rightarrow Q$	A
2.	$Q \rightarrow R$	A
3.	P	A
4.	Q	$1, 3 \rightarrow E$
5.	R	$2, 4 \rightarrow E$

We list the three premises first, number each line, and add the label "A" to indicate that each is an assumption. We then deduce the conclusion R via two steps of reasoning (both with the same form): from a conditional and its antecedent we may infer its consequent. This rule is known as *conditional elimination* or *modus ponens* and is denoted by $\rightarrow$ E. The first step is from lines 1 and 3 to line 4; the second is from lines 2 and 4 to 5. The numbers to the right denote the lines from which 4 and 5 are derived and $\rightarrow$ E denotes the rule of inference.

We list eight of the ten basic rules of inference and then consider some examples highlighting the possibilities and limitations of these rules.

> *Negation Elimination* (~ E): From a wff ~ ~ φ, we infer φ.
>
> *Conditional Elimination* (→ E): From wffs φ → ψ and φ, we infer ψ (also known as *modus ponens*).
>
> *Conjunction Introduction* (& I): From wffs φ and ψ, we infer the conjunction φ & ψ (also known as *conjunction*).
>
> *Conjunction Elimination* (& E): From a wff φ & ψ, we infer either φ or ψ (also known as *simplification*).
>
> *Disjunction Introduction* (∨ I): From a wff φ, we infer the disjunction of φ with any wff (also known as *addition*).
>
> *Disjunction Elimination* (∨ E): From wffs φ ∨ ψ, φ → χ, ψ → χ, we infer χ (also known as *constructive dilemma*).
>
> *Biconditional Introduction* (↔ I): From wffs φ → ψ and ψ → φ, we infer φ ↔ ψ.
>
> *Biconditional Elimination* (↔ E): From a wff φ ↔ ψ, we infer either φ → ψ or ψ → φ.

This list includes neither Negation Introduction nor Conditional Introduction; these *hypothetical rules of inference* are discussed in the next section. In addition, note that the rules say nothing about the complexity of the wffs to which they are applied. We use the Greek letters φ, ψ, and χ to indicate that the rules apply to all wffs, whether they are sentence letters or compound wffs.

We first consider Negation Elimination (~ E). If we begin with a doubly negated sentence (e.g., "It is not the case that Richard Nixon was not president") and remove both negations ("Richard Nixon was president"), the resulting sentence has the same truth value as the original. Thus the inference from a doubly negated sentence to the result of removing both negations is valid; consider the following:

Example 4.1. Prove: ~ P→ ~ ~ Q, ~ ~ ~ P ⊢ Q.

1.	~ P→ ~ ~ Q	A
2.	~ ~ ~ P	A
3.	~ P	2 ~ E
4.	~ ~ Q	1, 3 → E
5.	Q	4 ~ E

The rule ~E does *not* permit us to reason from line 1 to ~ $P \rightarrow Q$, since line 1 is a conditional, not a doubly negated wff; we must *detach* ~ ~ Q as in step 4 before applying ~ E.

The following derivation illustrates the use of both Conjunction Introduction (&I) and Conjunction Elimination (&E):

Example 4.2. Prove: $P \rightarrow (Q \& R)$, $P \vdash P \& Q$.

1. $P \rightarrow (Q \& R)$ A
2. P A
3. $Q \& R$ 1, 2 $\rightarrow$ E
4. Q 3 &E
5. $P \& Q$ 2, 4 &I

We drop the outer parentheses from $(Q \& R)$ in step 3, following our convention from Chapter 3.

The validity of Disjunction Introduction ($\vee$I) is an immediate consequence of the truth table for $\vee$. If either or both of the disjuncts are true, then the whole disjunction is true. For example, if the statement "Today is Tuesday" is true, then the disjunction "Today is either Tuesday or Wednesday" must also be true (as well as "Today is either Wednesday or Tuesday"). The following derivation uses Disjunction Introduction:

Example 4.3. Prove: P, ~ ~$(P \rightarrow Q) \vdash (R \& S) \vee Q$.

1. P A
2. ~ ~$(P \rightarrow Q)$ A
3. $P \rightarrow Q$ 2 ~E
4. Q 1, 3 $\rightarrow$ E
5. $(R \& S) \vee Q$ 4 $\vee$ I

According to Disjunction Elimination ($\vee$E), from wffs of the form $\varphi \vee \psi$, $\varphi \rightarrow \chi$, and $\psi \rightarrow \chi$, we may infer χ. To illustrate, assume that today is either Saturday or Sunday. In addition, assume that if it is Saturday, then tonight there will be a concert, and that if it is Sunday, then tonight there will be a concert. We can infer that tonight there will be a concert without knowing whether it is Saturday or Sunday. Consider the following derivation:

Example 4.4. Prove: $(P \lor Q) \& R, P \to S, Q \to S \vdash S \lor T.$

1. $(P \lor Q) \& R$	A
2. $P \to S$	A
3. $Q \to S$	A
4. $P \lor Q$	1 &E
5. S	2, 3, 4 $\lor$ E
6. $S \lor T$	5 $\lor$ I

Finally, the introduction and elimination rules for the biconditional function in a manner similar to those for conjunction.

Example 4.5. Prove: $P \leftrightarrow (Q \lor R), R \vdash P.$

1. $P \leftrightarrow (Q \lor R)$	A
2. R	A
3. $(Q \lor R) \to P$	1 $\leftrightarrow$ E
4. $Q \lor R$	2 $\lor$ I
5. P	3, 4 $\to$ E

Example 4.6. Prove: $P \to Q, (P \to Q) \to (Q \to P) \vdash P \leftrightarrow Q.$

1. $P \to Q$	A
2. $(P \to Q) \to (Q \to P)$	A
3. $Q \to P$	1, 2 $\to$ E
4. $P \leftrightarrow Q$	1, 3 $\leftrightarrow$ I

Hypothetical Inference Rules

The introduction rules for conditionals and negation differ from the other introduction rules in that they employ *hypothetical reasoning*. A *hypothesis* is a temporary assumption made "for the sake of argument" in order to show that a particular conclusion will follow.

For example, suppose a runner has injured her ankle a week before a big race. We would like to persuade her to stop running for a few days to let her ankle heal and so assert the conditional "If you keep running now, you won't be able to run the race." The most widely applicable way to prove a conditional is to hypothesize its antecedent (i.e., to assume it for the sake of argument) and then show its consequent must follow. Consider the following:

> Suppose you keep running on your swollen ankle. If you keep running, it will not heal in a week. If it does not heal in a week, then you won't be able to run the race. Thus, you won't be able to run the race.

In this hypothetical argument, the word "suppose" is an indicator signaling that "You keep running" is a hypothesis. In all, this argument employs the following three assumptions:

> Your ankle is swollen.
> If your ankle is swollen and you keep running, then it will not heal in a week.
> If your ankle does not heal in a week, then you won't be able to run in the race.

These assumptions serve as premises in our argument. We assert them to be true—unlike the hypothesis, which we assume only for the sake of argument. Given these premises, the hypothetical argument shows that *if* the hypothesis is true, then the conclusion must also be true. In the above example, there is no situation in which "You keep running" is true while "You won't be able to run the race" is false. Thus, the conditional "If you keep running now, then you won't be able to run the race" must be true. Formalizing the above argument, we obtain:

1. S A
2. $(S \,\&\, K) \to \sim W$ A
3. $\sim W \to \sim R$ A
4. K H
5. $S \,\&\, K$ 1, 4 & I
6. $\sim W$ 2, 5 $\to$ E
7. $\sim R$ 3, 6 $\to$ E
8. $K \to \sim R$ 4 – 7 $\to$ I

The assumptions are listed first and labeled "A." The hypothesis K is introduced at step 4 and labeled "H." To the left of K we begin a vertical line that extends downward to indicate the part of the proof in which the hypothesis is assumed; the hypothesis is said to be *in effect* for this marked portion of the proof. In the final step, in which we conclude $K \to \sim R$, we cite lines 4 to 7 (the hypothetical derivation) and the conditional implication rule ($\to$I). The formal statement of this rule is:

> *Conditional Introduction* ($\to$ I): Given a derivation of a wff ψ from a hypothesis φ, we discharge the hypothesis and we infer $\varphi \to \psi$ (also known as *conditional proof*).

Example 4.7. Prove: $P \to Q, Q \to R \vdash P \to R$.

1.	$P \to Q$	A
2.	$Q \to R$	A
3.	$\quad$ P	H
4.	$\quad$ Q	$1, 3 \to$ E
5.	$\quad$ R	$2, 4 \to$ E
6.	$P \to R$	$3 - 5 \to$ I

Hypothetical rules may be used more than once in the course of a derivation. In addition, since Conditional Introduction ($\to$I) is the primary strategy for proving conditionals and since the application of Disjunction Elimination ($\vee$E) requires two conditional premises, $\to$ I is often a precursor to $\vee$E; consider the following derivation:

Example 4.8. Prove: $P \vee Q \vdash Q \vee P$.

1.	$P \vee Q$	A
2.	$\quad$ P	H
3.	$\quad$ $Q \vee P$	$2 \vee$ I
4.	$P \to (Q \vee P)$	$2 - 3 \to$ I
5.	$\quad$ Q	H
6.	$\quad$ $Q \vee P$	$5 \vee$ I
7.	$Q \to (Q \vee P)$	$5 - 6 \to$ I
8.	$Q \vee P$	$1, 4, 7 \vee$ E

Sometimes we need to embed one hypothetical argument within another; for example, when a conclusion is a conditional with a conditional consequent. Consider the following derivation:

Example 4.9. Prove: $(P \& Q) \to R \vdash P \to (Q \to R)$.

1. $(P \& Q) \to R$			A
2.	P		H
3.		Q	H
4.		$P \& Q$	2, 3 & I
5.		R	1, 4 $\to$ E
6.	$Q \to R$		3 – 5 $\to$ I
7. $P \to (Q \to R)$			2 – 6 $\to$ I

Several important guidelines should be observed when using hypothetical reasoning:

1. Every hypothesis in a proof begins a new vertical line.
2. No occurrence of a wff to the right of a vertical line may be cited in any rule applied after that vertical line has ended.
3. If two or more hypotheses are in effect simultaneously, then the order in which they are discharged must be the reverse of the order in which they are introduced.
4. A proof is not complete until all hypotheses have been discharged.

The second hypothetical rule of inference is Negation Introduction (~ I). To prove a negated conclusion using this rule, we hypothesize the conclusion without its negation sign and derive from it an "absurdity" or contradiction of the form $\varphi \ \& \sim \varphi$ (that is, a conjunction whose second conjunct is the negation of the first). The chief distinction of contradictions is that they are always false. If we validly derive a false conclusion from a hypothesis, the hypothesis must be false and so the negation of the hypothesis must be true. Formally, we have:

Negation Introduction (~ I): Given a derivation of an absurdity from a hypothesis φ, we may discharge the hypothesis and we infer $\sim \varphi$.

Example 4.10. Prove: $P \to Q, \sim Q \vdash \sim P$.

1. $P \to Q$		A
2. $\sim Q$		A
3.	P	H
4.	Q	1, 3 $\to$ E
5.	$Q \ \& \sim Q$	2, 4 & I
6. $\sim P$		3 – 6 ~ I

Note!

When several hypotheses are in effect at once, we make the proof easier to read by appropriately labeling the hypotheses.

Example 4.11. Prove: $\sim P \vee \sim Q \vdash \sim (P \& Q)$.

1.	$\sim P \vee \sim Q$	A
2.	$\sim P$	H (for $\rightarrow$ I)
3.	$P \& Q$	H (for $\sim$ I)
4.	P	3 & E
5.	$P \& \sim P$	2, 4 & I
6.	$\sim (P \& Q)$	3 – 5 $\sim$ I
7.	$\sim P \rightarrow \sim (P \& Q)$	2 – 6 $\rightarrow$ I
8.	$\sim Q$	H (for $\rightarrow$ I)
9.	$P \& Q$	H (for $\sim$ I)
10.	Q	9 & E
11.	$Q \& \sim Q$	8, 10 & I
12.	$\sim (P \& Q)$	9 – 11 $\sim$ I
13.	$\sim Q \rightarrow \sim (P \& Q)$	8 – 12 $\rightarrow$ I
14.	$\sim (P \& Q)$	1, 7, 13 $\vee$E

Derived Rules

Every instance of a valid argument form is valid. Thus in proving the form $P \rightarrow Q, \sim Q \vdash \sim P$ (Example 4.10), we established the validity of every argument resulting from that form by uniformly replacing P and Q with sentences (no matter how complex). Consider:

> If it is raining or snowing, then the sky is not clear.
> It is not the case that the sky is not clear.
> $\therefore$ It is not either raining or snowing.

In this case P stands for "It is raining or snowing" and Q stands for "The sky is not clear". If we formalize this argument to reveal all of its logical structure, we obtain the following argument form:

$(R \vee S) \rightarrow {\sim}C, {\sim} {\sim}C \vdash {\sim}(R \vee S)$

This is a *substitution instance* of the original form; that is, it is the result of replacing zero or more of its sentence letters by wffs, each occurrence of the same sentence letter being replaced by the same wff. In this example, we replaced each P by $(R \vee S)$ and each Q by ${\sim}C$.

Inference rules that are derived from previously proved argument forms are called *derived rules*. For each previously proved form, the *associated inference rule* (or *associated derived rule*) is:

From the premises of any substitution instance of the form, we may validly infer the conclusion of that substitution instance.

Any previously proved form may be used as a derived rule in a proof. As justification, we cite the lines used as premises and the name of the derived rule (if it has one).

Example 4.12. Prove the derived rule Contradiction: $P, {\sim}P \vdash Q$.

1.	P	A
2.	${\sim}P$	A
3.	$\quad {\sim}Q$	H
4.	$\quad\ P \mathbin{\&} {\sim}P$	1, 2 & I
5.	${\sim}{\sim}Q$	3 – 4 ${\sim}$ I
6.	Q	5 ${\sim}$ E

Example 4.13. Prove the derived rule Disjunctive Syllogism: $P \vee Q, {\sim}P \vdash Q$.

1.	$P \vee Q$	A
2.	${\sim}P$	A
3.	$\quad P$	H (for $\rightarrow$ I)
4.	$\quad Q$	2, 3 CON
5.	$P \rightarrow Q$	3 – 4 $\rightarrow$ I
6.	$\quad Q$	H (for $\rightarrow$ I)
7.	$Q \rightarrow Q$	6 – 6 $\rightarrow$ I
8.	Q	1, 5, 7 $\vee$ E

As suggested by these examples, many derived rules have names. The following is a list of some important derived rules:

Modus Tollens (MT) From wffs φ → ψ and ~ ψ, we infer ~ φ.

Hypothetical Syllogism (HS) From wffs φ → ψ and ψ → χ, we infer φ → χ.

Absorption (ABS) From a wff φ → ψ, we infer φ → (φ & ψ).

Constructive Dilemma (CD) From wffs φ ∨ ψ, φ → χ, and ψ → ω, we infer χ ∨ ω.

Repeat (RE) From a wff φ, we infer φ.

Contradiction (CON) From wffs φ and ~ φ, we infer any wff.

Disjunctive Syllogism (DS) From wffs φ ∨ ψ and ~ φ, we infer ψ.

As the next two derived rules suggest, the theorems discussed in the following sections can also be used to make proofs shorter and more efficient.

Theorem Introduction (TI) Any substitution instance of a theorem may be introduced at any line of a proof.

Equivalence Introduction If φ ↔ ψ is a theorem and φ is a subwff of some wff χ, from χ we infer the result of replacing one or more occurrences of φ in χ by ψ.

Theorems

Some wffs are provable without any premises and are the *theorems* (or *laws*) of the propositional calculus. We write the turnstile symbol ⊢ in front of a wff to indicate that it is a theorem. In general, this symbol asserts that the wff on its own right is provable using only the wffs on its left as assumptions. Thus, when no wffs appear on its left, the formula on its right is a theorem.

Example 4.14. Prove the theorem: ⊢ P → (P ∨ Q).

1.	P	H
2.	P ∨ Q	1 ∨ I
3.	P → (P ∨ Q)	1 – 2 → I

Example 4.15. Prove the theorem: ⊢ P ∨ ~ P.

1.	~(P ∨ ~P)	H (for ~I)
2.	P	H (for ~I)
3.	P ∨ ~P	2 ∨ I
4.	(P ∨ ~P) & ~(P ∨ ~P)	1, 3 &I
5.	~P	2 – 4 ~I
6.	P ∨ ~P	5 ∨ I
7.	(P ∨ ~P) & ~(P ∨~P)	1, 6 &I
8.	~ ~(P ∨ ~P)	1 – 7 ~I
9.	P ∨ ~P	8 ~E

Equivalences

A biconditional that is a theorem is called an *equivalence*. If φ ↔ ψ is an equivalence, then φ and ψ validly imply one another and are said to be *interderivable*.

Example 4.16. Prove the equivalence: ⊢ (P → Q) ↔ ~ (P& ~Q).

1.	P → Q	H (for → I)
2.	P& ~Q	H (for ~I)
3.	P	2 &E
4.	Q	1, 3 → E
5.	~Q	2 &E
6.	Q& ~Q	4, 5 &I
7.	~(P& ~ Q)	2 – 6 ~I
8.	(P → Q) → ~(P& ~Q)	1 – 7 → I
9.	~(P& ~Q)	H (for → I)
10.	P	H (for → I)
11.	~Q	H (for ~I)
12.	P& ~Q	10, 11 &I
13.	(P& ~Q) & ~(P& ~Q)	9, 12 &I
14.	~ ~Q	11 – 13 ~I
15.	Q	14 ~E
16.	P → Q	10 – 15 → I
17.	~(P& ~Q) → (P → Q)	9 – 16 → I
18.	(P → Q) ~(P&Q)	8, 17 ↔ I

Like derived rules, many equivalences have names. The following is a list of some important equivalences:

De Morgan's law (DM)	$\sim (P \ \& \ Q) \leftrightarrow (\sim P \lor \sim Q)$
De Morgan's law (DM)	$\sim (P \lor Q) \leftrightarrow (\sim P \ \& \sim Q)$
Commutation (COM)	$(P \lor Q) \leftrightarrow (Q \lor P)$
Commutation (COM)	$\sim(P \ \& \ Q) \leftrightarrow (Q \ \& \ P)$
Association (ASSOC)	$(P \lor (Q \lor R)) \leftrightarrow ((P \lor Q) \lor R)$
Association (ASSOC)	$(P \ \& \ (Q \ \& \ R)) \leftrightarrow ((P \ \& \ Q) \ \& \ R)$
Distribution (DISTR)	$(P \ \& \ (Q \lor R)) \leftrightarrow ((P \ \& \ Q) \lor (P \ \& \ R))$
Distribution (DISTR)	$(P \lor (Q \ \& \ R)) \leftrightarrow ((P \lor Q) \ \& \ (P \lor R))$
Double Negation (DN)	$P \leftrightarrow \sim \sim P$
Transposition (TRANS)	$(P \to Q) \leftrightarrow (\sim Q \to \sim P)$
Material Implication (MI)	$(P \to Q) \leftrightarrow (\sim P \lor Q)$
Exportation (EXP)	$((P \ \& \ Q) \to R) \leftrightarrow (P \to (Q \to R))$
Tautology (TAUT)	$P \leftrightarrow (P \ \& \ P)$
Tautology (TAUT)	$P \leftrightarrow (P \lor P)$

 Note!

If two wffs are interderivable, one may be validly substituted for any occurrence of the other, as either a whole wff or as a subwff of some larger wff.

For example, since DN shows P and $\sim \sim P$ are interderivable, it also guarantees that $(Q \to \sim \sim P)$ is interderivable with $(Q \to P)$. In addition, the proof of an equivalence also establishes the derivability of all substitution instances of that equivalence. Thus, DN also asserts the interderivability of Q and $\sim \sim Q$, of $S \ \& \sim R$ and $\sim \sim (S \ \& \sim R)$, and so on.

Example 4.17. Prove: $(P \leftrightarrow Q) \vdash \sim ((P \to Q) \to \sim (Q \to P))$.

1. $P \leftrightarrow Q$	A
2. $P \to Q$	1 $\leftrightarrow$ E
3. $Q \to P$	1 $\leftrightarrow$ E
4. $(P \to Q) \ \& \ (Q \to P)$	2, 3 & I
5. $\sim \sim ((P \to Q) \ \& \ (Q \to P))$	4 DN
6. $\sim (\sim (P \to Q) \lor \sim (Q \to P))$	5 DM
7. $\sim ((P \to Q) \to \sim (Q \to P))$	6 MI

Chapter 5
THE LOGIC OF CATEGORICAL STATEMENTS

IN THIS CHAPTER:

✔ *Categorical Statements*
✔ *Venn Diagrams*
✔ *Immediate Inferences*
✔ *Categorical Syllogisms*

Categorical Statements

The propositional logic of Chapters 3 and 4 focuses on the relations generated by logical operators such as the connectives "not," "and," "or," "if-then," and "if and only if." This chapter takes a first look at the logical relations generated by such expressions as "all," "some," and "no." The primary reason to move beyond propositional logic is that the validity of some arguments does not depend on connectives. For example, the following valid argument cannot be proven valid via propositional logic:

Some four-legged creatures are gnus.
All gnus are herbivores.
∴ Some four-legged creatures are herbivores.

From the viewpoint of propositional logic, these sentences have no internal structure. Indeed, if we try to formalize this argument in propositional logic, the best we can do is something like:

$$P, Q \vdash R.$$

This argument form is invalid, since any three sentences P, Q, and R with P and Q true and R false constitutes a counterexample.

However, from a more discerning perspective, each sentence in this argument does have an internal structure, that together yields a valid argument form. These structures consist of relations between terms occurring within each sentence. For example, we can represent the above argument form as follows:

> Some F are G.
> All G are H.
> $\therefore$ Some F are H.

In this argument form, the letters F, G, and H are placeholders for the *class terms* "four-legged creature", "gnu", and "herbivore". Class terms (also called *predicates*) denote collections of objects.

Example 5.1. Formalize the following valid argument:

> All logicians are philosophers.
> Some logicians are mathematicians.
> $\therefore$ Some mathematicians are philosophers.

Using L for "logicians", P for "philosophers", and M for "mathematicians", the argument form is:

> All L are P.
> Some L are M.
> $\therefore$ Some M are P.

The class terms used above are all simple nouns. Noun phrases, such as "blue thing" or "my friend", are also class terms. Adjectives and adjective phrases may function as class terms as well; the adjective "old" denotes the set of all old things and the phrase "in the northern hemisphere" denotes the set of all things located in the northern hemisphere. In addition, verbs and verb phrases may be regarded as class terms; "move" and "love Bill" denote, respectively, the set of all things that move and the set of all things that love Bill.

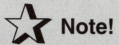

Note!

To facilitate substitution and comparison of class terms, we typically (re)write all class terms as nouns and noun phrases.

Example 5.2. Formalize the following valid argument:

All cups on the table have been painted by my art teacher.
All things painted by my art teacher are beautiful.
∴ All cups on the table are beautiful.

Using *C* for "cups on the table", *A* for "things painted by my art teacher", and *B* for "things that are beautiful" (note the conversion to noun phrases for *A* and *B*), the argument form is:

All *C* are *A*.
All *A* are *B*.
∴ All *C* are *B*.

Class terms are often related to one another by the *quantifiers* "all" and "some." Quantifiers are logical operators expressing relationships among the sets designated by class terms. For example, sentences of the form "All *A* are *B*" assert that the set *A* is a subset of *B*; that is, every member of *A* is also a member of *B*.

By a standard convention of logic, sentences of the form "Some *A* are *B*" assert that set *A* shares at least one member with set *B*. This departs from ordinary usage in two ways. First, "Some *A* are *B*" often signifies that set *A* shares *more than one* member with set *B*, whereas logic only requires *A* to share *at least one* member with *B*. Second, when we say "*Some A* are *B*," we often presume that not all *A* are *B*. In contrast, in the logical sense of "some," it is correct to say that some friends of mine are angry with me even when all of them are.

The only other expression occurring in the above arguments is "are." This word (together with its grammatical variants "is," "am," "was," etc.) is called the *copula*, because it couples or joins subject and predicate. Each sentence in our argument consists of a quantifier followed by a class term, a copula, and another class term. We call such sentences (and their negations) *categorical statements*.

In a categorical statement, the first class term is the *subject term* and the second class term is the *predicate term*. Unnegated categorical statements come in four distinct forms each denoted by a vowel (here S stands for the subject term and P for the predicate term):

Designation	Form
A	All S are P.
E	No S are P.
I	Some S are P.
O	Some S are not P

Observe that O-form statements contain the expression "not," which plays two distinct roles in categorical statements. When applied to an entire sentence, "not" expresses negation: the logical operation that makes true sentences false and false sentences true. In contrast, when "not" modifies only a class term, the result is a new class term. For example, in the O-form statement "Some trees are not oaks", "not" modifies only the class term "oaks" resulting in the new class term "not oaks" or "non-oaks" designating the set of all things that are not oak trees. In general, the set of all things that are not members of a given set S is called the *complement* of S and when "not" is applied to a class term, it is said to express *complementation* rather than negation.

Important!

Since "not" has a dual role in categorical statements, the reader must decide if "not" applies to the entire sentence (expressing negation) or an individual class term (expressing complementation).

Example 5.3. Analyze the logical form of the following categorical statement: All men are not rational.

The analysis of this sentence depends on whether "not" is read as negation or as complementation. If read as negation, then (using M for "men" and R for "rational things") the argument form is "~ (All M are R)," a negated A-form statement asserting that not all men are rational but leaving open the possibility that some are. If "not" is read as complementation, then the argument form is "All M are non-R," an A-form statement asserting that all men are nonrational.

The prefix "non-" expresses complementation unambiguously in categorical statements. Prefixes, such as "un-," "im-," "in-," and "ir-" may express complementation, but they can also express other forms of opposition. For example, the set of unhappy things is not the complement of the set of happy things. Many things, such as rocks, are neither happy nor unhappy. The complement of the set of happy things is the set of non-happy things (which include things that are unhappy as well as things that are neither happy nor unhappy).

The class terms of a categorical statement may occur with or without complementation operators. Sentences of the form "Some S are not P" actually have two forms; we write "Some S are not P" when the O-form is to be emphasized and "Some S are non-P" when the I-form is to be emphasized.

If more than one complementation operator is applied to a single class term, the two negatives cancel out (so, double complementation behaves like double negation). For example, "All men are not nonmortal" is logically equivalent to "All men are mortal". In contrast, the sentence "It is not the case that all men are not mortal" involves one negation and one complementation, and they cannot be cancelled without altering the meaning of the sentence.

Example 5.4. Formalize the following sentences and, if the sentence is a categorical statement, specify its form.

  (a) All embezzlers are wicked.
  (b) Not all embezzlers are wicked.
  (c) Some embezzlers are not wicked.
  (d) If Jack is an embezzler, then Jack is wicked.
  (e) Nobody in this room is leaving.
  (f) Someone in this room is leaving and someone isn't.

(a) All *E* are *W* is an A-form statement.
(b) ~ (All *E* are *W*) is the negation of an A-form statement.
(c) Some *E* are not *W* is an O-form statement.
(d) *J* → *K* is a conditional statement, not categorical.
(e) No *P* are *L* is an E-form statement.
(f) Some *P* are *L* & Some *P* are not *L* is a conjunction of an I-form statement and an O-form statement, but is not itself a categorical statement.

Venn Diagrams

In working with categorical statements, it often helps to visualize relationships among sets using *Venn diagrams.* We represent a categorical statement in a Venn diagram with two overlapping circles corresponding to the sets designated by the statement's subject and predicate terms. The area inside a circle represents the contents of the set; the area outside it represents the complement. The area of overlap between the two circles represents the members (if any) shared by the corresponding sets.

We show that a set or part of a set has no members by shading the representative portion of the Venn diagram. For example, the E-form statement "No *S* are *P*" asserts that sets *S* and *P* share no members. The corresponding Venn diagram consists of two overlapping circles (one for *S* and one for *P*) with the overlapping part shaded.

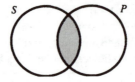

Figure 5-1

Similarly, the A-form statement "All *S* are *P*" says that *S* is contained in *P*, so that *S* has no members that are not also in *P*. Thus, in its Venn diagram, the part of the *S* circle outside the *P* circle is shaded.

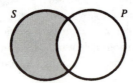

Figure 5-2

We show a set or part of a set has at least one member by putting an X in the representative part of the Venn diagram. For example, the I-form statement "Some S are P" says that S and P share at least one member and we place an X in the overlapping part of the two circles.

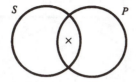

Figure 5-3

Similarly, the O-form categorical statement "Some S are not P" asserts that S has at least one member not in P and so we place an X in the part of the S circle outside of the P circle.

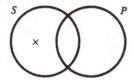

Figure 5-4

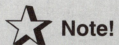

⭐ Note!

If part of a Venn diagram is not shaded and does not contain an X, we do not know whether or not the area contains members.

For statements involving complementation we frame the diagram. The members of a set's complement are represented by the area outside the set's circle but inside the frame.

Example 5.5. Draw a Venn diagram for "All non-S are P".

This form asserts that the complement of *S* is contained in *P*; that is, the complement of *S* outside of *P* is empty. In the Venn diagram, the complement of *S* is represented by the area outside the *S* circle but inside the frame, so we shade this part of the diagram except for the portion corresponding to the P circle.

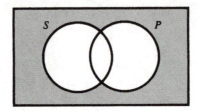

Figure 5-5

Example 5.6. Draw a Venn diagram for "Some non-*S* are not *P*".

This form says that some members of the complement of *S* are also members of the complement of *P*. The complement of a set is represented by the area outside its circle but inside the frame. Therefore, we place an X outside both circles but inside the frame.

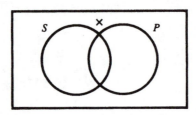

Figure 5-6

You Should Know This

We represent negation on a Venn diagram by swapping shadings and X's; that is, shade any area with an X and replace any shading with an X.

Example 5.7. Draw a Venn diagram for "~ (All S are P)".

Figure 5-2 has the Venn diagram for "All S are P" (the A-form). We diagram its negation by replacing the shading with an X.

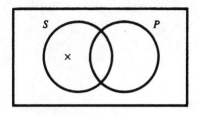

Figure 5-7

The Venn diagram for the negated A-form "~ (All S are P)" from Example 5.7 is the same as the diagram for the O-form "Some S are not P" given in Figure 5-4. This shows that "~ (All S are P)" and "Some S are not P" are two different ways of saying the same thing, and so are logically equivalent. Similarly, one can show that the negated O-form "~ (Some S are not P)" and the A-form "All S are P" have the same Venn diagrams and so are logically equivalent. These observations play an important role in the next section on immediate inferences.

Immediate Inferences

Immediate inferences are one-premise arguments whose premise and conclusion are both categorical statements. We will consider four types of immediate inferences: contradictories, conversion, contraposition, and obversion. And, we will test the validity of immediate inference forms via Venn diagrams. The first step is to diagram the premise. If the resulting diagram is also the diagram of the conclusion, then the immediate inference form is valid; if not, the form is invalid.

Important!

If two statement forms have exactly the same Venn diagram, they are logically equivalent, and we can make an immediate inference from one to the other.

Beginning with contradictories, recall from Example 5.7 that "~ (All *S* are *P*)" and "Some *S* are not *P*" have the same Venn diagram, and so are logically equivalent. In addition, we observed that "All *S* are *P*" and "~ (Some *S* are not *P*)" are logically equivalent. Taken together, these results show that any pair of A-form and O-form statements with the same subject and predicate terms are *contradictories*; that is, each validly implies the negation of the other and the negation of each validly implies the other. Similarly, two I-form and E-form statements with the same subject and predicate terms are contradictories. One valid immediate inference is from one of a pair of contradictories to the negation of the other, or vice versa.

Continuing our study, two categorical statements are said to be *converses* if one results from the other by exchanging subject and predicate terms. For example, "Some *P* are *S*" is the converse of "Some *S* are *P*". An inference from a categorical statement to its converse is called *conversion* and is a valid immediate inference for forms E and I, but is invalid for forms A and O.

Example 5.8. Use a Venn diagram to show that conversion is valid for E-form statements "No *S* are *P*".

We compare the Venn diagram for the E-form given in Figure 5-1 with the following diagram (Figure 5-8) of its converse "No *P* are *S*". These diagrams are identical since both assert that the sets *S* and *P* share no members. Therefore, an E-form statement and its converse are logically equivalent and conversion is valid for E-form statements.

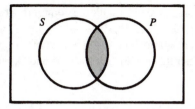

Figure 5-8

Example 5.9. Use a Venn diagram to show that conversion is not valid for A-form statements "All *S* are *P*".

We compare the Venn diagram for the A-form given in Figure 5-2 with the following diagram (Figure 5-9) of its converse "All *P* are *S*". The diagram for "All *S* are *P*" indicates that the set of things that are *S* and not *P* is empty, while the diagram for "All *P* are *S*" indicates that the set of things that are *P* and not *S* is empty. Since the shadings are different, conversion is invalid for A-form statements.

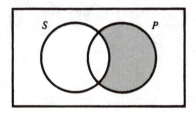

Figure 5-9

The following is a specific counterexample witnessing the invalidity of conversion for A-form statements:

All salesmen are people.
∴ All people are salesmen.

Two categorical statements are *contrapositives* if one results from the other by replacing its subject term with the complement of its predicate term and its predicate term with the complement of its subject term. For example, "All *S* are *P*" and "All non-*P* are non-*S*" are contrapositives. *Contraposition* is an immediate inference from one of a pair of contra-positives to the other that is valid for forms A and O, but invalid for forms E and I.

Example 5.10. Use a Venn diagram to show that contraposition is not valid for E-form statements "No *S* are *P*".

We compare the Venn diagram for the E-form given in Figure 5-1 with the following diagram of its contrapositive "No non-*P* are non-*S*". The contrapositive asserts that the complement of *P* and the complement of *S* share no members, and so we shade the region corresponding to things that are both non-*S* and non-*P*. Since the shadings are different, contraposition is invalid for E-form statements

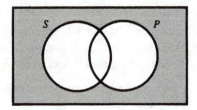

Figure 5-10

The following is a specific counterexample witnessing the invalidity of contraposition for E-form statements:

No snakes are people.
∴No non-people are non-snakes.

Before discussing our fourth type of immediate inference, we note that the four basic forms of categorical statements may be classified by *quantity* and *quality*. The two quantities are *universal* and *particular*, and the two qualities are *affirmative* and *negative*. The four basic forms have the following characteristic combinations of quality and quantity:

Form	Quantity	Quality
A	Universal	Affirmative
E	Universal	Negative
I	Particular	Affirmative
O	Particular	Negative

Our final type of immediate inference involves changing the quality of a categorical statement (while keeping its quantity the same) and replacing the predicate term by its complement. Such an inference is called *obversion*, and statements obtained from one another by obversion are called *obverses*. For example, the obverse of "No *S* are *P*" is "All *S* are non-*P*". Obverse categorical statements always have identical Venn diagrams, and so are always logically equivalent.

Example 5.11. Use a Venn diagram to show that obversion is valid for A-form statements "All *S* are *P*".

The Venn diagram for the A-form given in Figure 5-2 is identical with the following diagram of its obverse "No *S* are non-*P*". Therefore, obversion is valid for A-form statements.

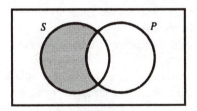

Figure 5-11

The logic originally developed by Aristotle for categorical statements identified some immediate inferences as valid that are not considered valid in modern logic. This discrepancy is due to the presupposition in Aristotelian logic that all subject and predicate terms designate nonempty sets (have at least one member), while modern logic does not make this assumption. Indeed, the presupposition of nonemptiness limited the applicability of Aristotelian logic. Dropping this assumption greatly simplified logic and resulted in dropping the extra inferences.

However, the acceptance of empty terms by modern logic created a problem: what is the truth value of A-form statements with empty subject terms? For example, is the sentence "All submarines over a mile long are pink" true or false? Modern logic stipulates that *every* A-form statement with an empty subject term is true. Thus, "All submarines over a mile long are pink" is true, as is "All submarines over a mile long are not pink". Notice that the "not" in this second sentence expresses complementation, rather than negation, and so both sentences being true is not a contradiction.

Categorical Syllogisms

Categorical syllogisms are two-premise arguments consisting of categorical statements with exactly three class terms: the subject and predicate term of the conclusion (these are, respectively, the *minor* and *major* terms of the syllogism) and a third term (the *middle* term) which occurs in both premises. In addition, the major and minor terms must each occur exactly once in a premise. For example, the following argument from the beginning of the chapter is a categorical syllogism:

> Some four-legged creatures are gnus.
> All gnus are herbivores.
> ∴ Some four-legged creatures are herbivores.

In this syllogism, the major term is "herbivores", the minor term is "four-legged creatures", and the middle term is "gnus".

We test for the validity of categorical syllogisms with Venn diagrams. The diagram of a syllogistic form uses three overlapping circles to represent the three terms in the premises. The circles are labeled (in any order) with letters standing for these terms. For the above syllogism, we use H for "herbivores", F for "four-legged creatures", and G for "gnus" to obtain the following:

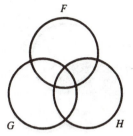

Figure 5-12

 **Note!**

The Venn diagram for a categorical syllogism must include a three-cornered middle region to represent the things common to all three sets.

We diagram one premise at a time. The resulting diagram may then be used to test the form for validity in the same way we tested immediate inferences: if the diagram of the premises is the same as the diagram of the conclusion, then the form is valid; if not, the form is invalid.

Example 5.12. Use a Venn diagram to test the validity of the following categorical syllogism:

No *F* are *G*.
All *G* are *H*.
∴ No *F* are *H*.

The first premise asserts that the sets *F* and *G* share no members, and so we shade the lens-shaped area shared by the *F* and *G* circles. The second premise says that *G* is a subset of *H*, and so we shade the crescent part of the *G* circle opposite *H*.

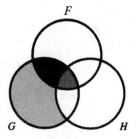

Figure 5-13

We have no information about the region of overlap between the F and H circles, which is consistent with the premises that there are Fs which are in H but not in G. If there is such an F, then the conclusion "No F are H" is false, and so the conclusion may be false while the premises are true. Thus, this form is invalid.

Example 5.13. Use a Venn diagram to test the validity of the following categorical syllogism:

All F are G.
No G are H.
∴No F are H.

Via the same approach as above, we produce the following Venn diagram for the two premises:

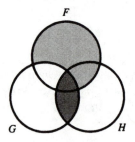

Figure 5-14

These premises shade the lens-shaped area shared by the *F* and *H* circles, showing that the conclusion "No *F* are *H*" must be true when the premises are true. Therefore, the form is valid.

Chapter 6
PREDICATE LOGIC

IN THIS CHAPTER:

- ✔ *Quantifiers and Variables*
- ✔ *Predicates and Names*
- ✔ *Formation Rules*
- ✔ *Models*
- ✔ *Refutation Trees*
- ✔ *Identity*

Quantifiers and Variables

In this chapter, we study *predicate logic*, which combines quantifiers ("all," "some," and "no") with the operators of propositional logic (the connectives). This apparatus is very powerful and is adequate for testing the validity of a wide range of argument forms.

We begin with a reformulation of categorical statements that reveals the presence of the logical operators from Chapters 3 and 4. For example, the A-form statement "All *S* are *P*" is logically equivalent to "For all *x*, if *x* is *S*, then *x* is *P*"; this reformulation contains the "if . . . then" conditional operator. Similarly, the E-form statement "No *S* are *P*" can be reformulated as "For all *x*, if *x* is *S*,

then it is not the case that x is P", which contains both a negation and a conditional.

In order to formalize such sentences, we adopt the *universal quantifier* symbol "∀" to mean "for all," "for any," and the like, and we symbolize "x is S" with "Sx". In addition, we use the symbolic notation → for the conditional and ~ for the negation. Using this notation, we formalize the reformulation of the A-form statement "All S are P" as ∀ $x(Sx → Px)$ and the reformulation of the E-form statement "No S are P" as ∀$x(Sx → {\sim}Px)$.

To formalize the I-form statement "Some S are P" and the O-form statement "Some S are not P", we adopt the *existential quantifier* symbol "∃" to mean "for some," "for at least one," "there exists . . . such that," and the like; as in the previous chapter, we shall understand "some" to include the limiting cases of "exactly one" and "every" object in the domain. Following our work above, we reformulate the I-form statement "Some S are P" as "For some x, x is S and x is P", which is formalized as ∃ $x(Sx \ \& \ Px)$. Similarly, we reformulate the O-form statement "Some S are not P" as "For some x, x is S and x is not P", which is formalized as ∃ $x(Sx \ \& \ {\sim}Px)$.

Notice that formalizing the I-form statement "Some S are P" as ∃ $x(Sx → Px)$ is *wrong*. This incorrect formalization means "There exists an x such that, if x is S, then x is P", which can be true even if nothing is S. Similarly, we cannot formalize an O-form statement with a conditional. In contrast, A-form and E-form statements *must* be formalized using conditionals, rather than conjunctions. For example, formalizing the A-form statement "All S are P" as ∀ $x(Sx \ \& \ Px)$ is *wrong*. This incorrect formalization can be interpreted as "For all x, x is both a shark and a predator" (that is, everything is a predatory shark), which is *not* logically equivalent to "All sharks are predators".

Example 6.1. Interpreting the letter F as "is a frog" and G as "is green", formalize the following sentences.

(a) Frogs are green.
(b) Some frogs are not green.
(c) Frogs are not green.
(d) Not everything is a frog.
(e) Nothing is a frog.
(f) Only frogs are green.

(a) ∀ x (Fx → Gx). This formulation treats "frogs" as expressing a universal quantification as is customary in predicate logic.

(b) ∃ x (Fx & ~Gx).This formalization treats "not" as expressing negation rather than complementation.

(c) Either of ∀ x (Fx → ~Gx) or ~ ∃ x (Fx & Gx) is correct.

(d) ~∀ x Fx

(e) Either of ∀x ~Fx or ~∃ x Fx is correct.

(f) ∀ x (Gx → Fx)

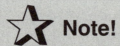 **Note!**

We must be careful in formalizing quantifiers since there is no direct, uniform match of English phrases with the symbols ∀ and ∃.

For example, "something" typically indicates an existential quantifier. However, the sentence "If something is cheap, then it is not of good quality" is a general statement to the effect that cheap things are not of good quality, and in this context, the word "something" expresses a universal quantification. Also, quantification is often not expressed by an explicit quantifier phrase, but by an adverb (such as "always" or "sometimes") or by some other word. For example, the statement "Only expensive things are nice" is an A-form statement "All *N* are *E*".

Example 6.2. Interpreting the letter *F* as "is a frog" and *G* as "is green", formalize the following sentences.

(a) If anything is green, then frogs are green.
(b) If everything is green, then frogs are green.
(c) Occasionally, frogs are green.

(a) ∃ x Gx → ∀ x (Fx → Gx)
(b) ∀ x Gx → ∀ x (Fx → Gx)
(c) ∃ x (Fx & Gx)

Predicates and Names

Not all sentences contain quantifiers. For example, a simple subject-predicate sentences, such as "Jones is a thief", attributes a property to an individual. We interpret lowercase letters from the beginning of the alphabet as names of individuals, and we adopt the convention (contrary to English grammar) of writing the subject after the predicate. Interpreting *j* as the name "Jones" and *T* as the predicate "is a thief", we formalize "Jones is a thief" as *Tj*. Note that the sentence "Jones is a thief" is true if Jones really is a thief, and false otherwise.

Important!

Combining a predicate with a name always yields a statement, which is true or false depending on whether or not the object designated by the name is a member of the class designated by the predicate.

Some *relational predicates* designate relationships between two or more objects and are combined with two or more proper names to yield a statement. These are usually written in logical notation in the order predicate-subject-object. For example, the statement "Bob loves Cathy" is formalized as *Lbc* and the statement "Cathy loves Bob" is formalized as *Lcb*. A predicate which takes only one name is called a *nonrelational* or *one-place* predicate. Similarly, a relational predicate that takes two names is called a *two-place* predicate; one that takes three names is called a *three-place* predicate; and so on.

Example 6.3. Interpreting the letters *a*, *b*, and *c* as "Alex", "Bob" and "Cathy", *M* as the one-place predicate "is a mechanic", *L* as the two-place predicate "likes", and *I* as the three-place predicate "introduced . . . to", formalize the following sentences.

(a) Cathy and Bob are mechanics.
(b) Bob likes Cathy.
(c) Bob likes himself.
(d) Cathy likes either Bob or Alex.
(e) Cathy introduced herself to Bob but not Alex.

(a) *Mc & Mb*
(b) *Lbc*
(c) *Lbb*
(d) *Lcb* ∨ *Lca*
(e) *Iccb & ~ Icca*

Some sentences involve both quantifiers and names. For example, the sentence "Jones likes everything" combines the predicate "likes" with the name "Jones" and the quantifier "everything". In such cases, the quantifier acts as a logical operator (rather than as a name). In particular, we reformulate the sentence "Jones likes everything" as "For all x, Jones likes x", which is formalized as ∀x Ljx. Sentences involving more than one quantifier are treated similarly. For example, we reformulate "Something likes everything" as "There exists an x such that, for all y, x likes y", which is formalized as ∃x∀y Lxy.

Example 6.4. Interpreting the letter b as "Bob" and L as the two-place predicate "likes", formalize the following sentences.

(a) Bob likes nothing.
(b) Nothing likes Bob.
(c) Something likes itself.
(d) If Bob likes himself, then he likes something.
(e) If Bob doesn't like himself, then he likes nothing.
(f) Everyone likes at least one thing.

(a) ∀ x ~Lbx
(b) ∀ x ~Lxb
(c) ∃ x Lxx
(d) Lbb → ∃x Lbx
(e) ~Lbb → ∀x ~Lbx
(f) ∀ x ∃ y Lxy

Some English sentences that combine a predicate with one or more nouns are formalized by introducing quantified variables. For example, the sentence "Jones likes a mechanic" is reformulated as "There exists an *x* such that *x* is a mechanic and Jones likes *x*" and is formalized as ∃ *x(Mx & Ljx)*. If a sentence involves more than one noun, the formalization will involve a corresponding number of quantified variables. For example, the sentence "A nurse likes a mechanic" is reformulated as "There exists a nurse *x* and a mechanic *y* such that *x* likes *y*" and is formalized as ∃ *x* ∃ *y ((Nx & My) & Lxy)*.

Example 6.5. Interpreting the letters *a* and *b* as "Alex" and "Bob"; *M* as the one-place predicate "is a mechanic"; *L* as the two-place predicate "likes"; and *I* as the three-place predicate "introduced . . . to", formalize the following sentences.

(a) A mechanic likes Bob.
(b) Every mechanic likes Bob.
(c) A mechanic introduced Bob to Alex.
(d) A mechanic introduced herself to Bob and Alex.

(a) ∃ *x (Mx & Lxb)*
(b) ∀ *x (Mx → Lxb)*
(c) ∃ *x (Mx & Ixba)*
(d) ∃ *x (Mx & (Ixxb & Ixxa))*

When formalizing sentences, keep in mind the following:

1. *Different variables do not necessarily designate different objects.* For example, ∀*x*∀*y Lxy* asserts that everything likes both itself and every other thing.
2. *The choice of variables makes no difference to meaning.* For example, ∃ *x Lxx* has the same meaning as both ∃ *y Lyy* and ∃ *z Lzz*. However, when two or more quantifiers govern overlapping parts of the same formula, different variables must be used. For example, ∀ *x* ∃ *y Lyx* and ∀ *x* ∃ *z Lzx* have the same meaning, but a different meaning than ∀ *x* ∃ *x Lxx*.
3. *The same variable used with two different quantifiers does not necessarily designate the same object in each case.* For example, "There is something Bob likes and something that Cathy likes"

can be formalized correctly as either ∃ *x Lbx* & ∃ *x Lcx* or
∃*x Lbx* & ∃ *y Lcy*.

4. *Many English sentences blending universal and existential quantifiers are ambiguous.* For example, the sentence "Everything likes something" can mean either of the non-equivalent sentences "There is some one thing that is liked by everything" or "Everything likes at least one thing" (which may not necessarily be the same thing in every case).

5. *The order of consecutive quantifiers only affects meaning when universal and existential quantifiers are mixed.* For example, ∃ *x* ∀ *y Lxy* and ∀ *y* ∃ *x Lxy* have different meanings, while both ∃ *x* ∃ *y Lxy* and ∃ *y* ∃ *x Lxy* mean "Something likes something" and are logically equivalent.

6. *Nested quantifiers may combine with truth-functional operators in many equivalent ways.* For example, the sentence "Cathy introduced a mechanic to a nurse" can be formalized as ∃ *x* ∃ *y((Mx* & *Ny)* & *Icxy)*, as ∃ *x (Mx* & ∃ *y(Ny* & *Icxy))*, and as well as other formalizations.

Formation Rules

As with propositional logic, we divide the vocabulary of the language of predicate logic into two parts: the *logical symbols* (whose interpretation or function remains fixed in all contexts) and the *nonlogical symbols* (whose interpretation varies from context to context):

Logical Symbols
Logical operators: ~, &, ∨, →, ↔
Quantifiers: ∀, ∃
Variables: lowercase letters "u" through "z"
Parentheses: (,)

Nonlogical Symbols
Name letters: lowercase letters "*a*" through "*t*"
Predicate letters: uppercase letters

We also permit the addition of numerical subscripts to these symbols. For example, a_3 is a name letter and P_{147} is a predicate letter.

We define a *formula* in our language as any finite sequence of logical and nonlogical symbols. An *atomic formula* is a predicate letter followed by zero or more name letters; these are used to formalize sentences when there is no need to represent their internal structure. A predicate letter followed by *n* name letters represents an *n-place predicate*. A well-formed formula, or wff, of predicate logic is defined via the following formation rules (again, using Greek letters to represent arbitrary expressions in our language):

Rule 1. Any atomic formula is a wff.

Rule 2. If φ is a wff, then so is ~ φ.

Rule 3. If φ and ψ are wffs, then so are (φ & ψ), (φ ∨ ψ), (φ → ψ), and (φ ↔ ψ).

Rule 4. If φ is a wff containing a name letter α, then any formula of the form ∀ β φ($^β/_α$) or ∃ β φ($^β/_α$) is a wff, where φ($^β/_α$) is the result of replacing one or more occurrences of α in φ by some variable β not already in φ.

Rule 5. Only formulas obtainable by (possibly) repeated applications of formation Rules 1 to 4 are wffs.

As in the propositional calculus, we adopt the informal convention of dropping paired outer parentheses.

Note that Rule 1 is broader than the corresponding rule of propositional logic in light of the definition of atomic formula. Rules 2 and 3 are the same as for propositional logic. However, Rule 4 is completely new. To illustrate, we let φ be the particular unquan- tified wff *(Fa & Gab)* and use Rule 4 to generate some quantified formulas. This wff contains two name letters *a* and *b*, either of which can serve as what Rule 4 calls α; we use *a* for this example. Rule 4 allows us to substitute β for α using some variable β not already in φ. Since φ contains no variables, any variable will work and we let β be the variable *x*. Three formulas result from replacing one or more occurrences of α = *a* in φ = *(Fa & Gab)* with β = *x*:

(Fx & Gxb) if both occurrences of *a* are replaced by *x*
(Fx & Gab) if only the first occurrence of *a* is replaced by *x*
(Fa & Gxb) if only the second occurrence of *a* is replaced by *x*

These formulas are not wffs, but Rule 4 stipulates that prefixing them with a quantifier followed by $\beta = x$ (that is, either $\forall \beta \, \varphi(^\beta/_\alpha)$) or $\exists \beta \, \varphi(^\beta/_\alpha)$) is a wff. Thus, all of the following are wffs:

$\forall x \, (Fx \, \& \, Gxb)$ $\forall x \, (Fx \, \& \, Gab)$ $\forall x \, (Fa\&Gxb)$
$\exists x \, (Fx \, \& \, Gxb)$ $\exists x \, (Fx \, \& \, Gab)$ $\exists x \, (Fa\&Gxb)$

Other wffs can be generated using the letter name b for α or variables other than x for β. For example, since $\forall x \, (Fx \, \& \, Gxb)$ is a wff, we can apply Rule 4 a second time using $\alpha = b$ and $\beta = y$ to obtain the wffs: $\forall y \, \forall x \, (Fx \, \& \, Gxy)$ and $\exists y \, \forall x \, (Fx \, \& \, Gxy)$.

 **Note!**

Rule 4 is the only rule allowing us to introduce variables into a wff, and introducing a variable requires us to prefix the formula with a quantifier for that variable. Thus, *Fx* and $\forall x \, Fa$ are not wffs.

Finally, in Rule 4, the clause "by some β not already in φ" ensures that quantifiers for the same variable never apply to overlapping parts of the same formula. For example, $\exists x \, Lxa$ is a wff by Rules 1 and 4, but we cannot apply Rule 4 again to obtain $\forall x \, \exists x \, Lxx$.

Example 6.6. Show that $\sim \exists x \, (\sim Fx \, \& \, \forall z \, Gzx)$ is a wff.

By Rule 1, *Fa* and *Gba* are wffs. By Rule 2, $\sim Fa$ is a wff and, by Rule 4, $\forall z \, Gza$ is a wff. Applying Rule 3 to these two wffs shows that $(\sim Fa \, \& \, \forall z \, Gza)$ is a wff, and then, by Rule 4, $\exists x \, (\sim Fx \, \& \, \forall z \, Gzx)$ is a wff. Finally $\sim\exists x \, (\sim Fx \, \& \, \forall z \, Gzx)$ is a wff by Rule 2.

Example 6.7. Explain why the following formulas are not wffs:

   (a) $\forall x \, Lxz$
   (b) $(\exists x \, Fx \, \& \, Gx)$
   (c) $\exists x \, \forall y \, Fx$
   (d) $\exists x \, \exists x \, (Fx \, \& \, \sim Gx)$

(a) The variable z is not quantified.
(b) The final occurrence of x is not quantified; for the sake of contrast, note that $\exists\, x\ (Fx\ \&\ Gx)$ is a wff.
(c) $\forall\, y$ requires another occurrence of y (see Rule 4).
(d) Quantifiers apply to overlapping parts of the formula.

Models

We now consider the semantics of predicate logic. There are two sources of complexity. One is that some atomic formulas of predicate logic are treated as compound expressions. These new atomic formulas (such as *Fa*) cannot be assigned a truth value directly; their truth value depends on the interpretation of the predicate and name letters occurring in the formula. Second, the new language contains new logical operators (the quantifiers) that are not part of propositional logic, and Venn diagrams cannot adequately deal with such linguistic structures.

We overcome the first difficulty with a *model*, or *interpretation structure*. A model specifies a universe together with a corresponding interpretation of the nonlogical symbols occurring in some given set of wffs. A *universe* or *domain* of interpretation is the class of objects relative to which the name letters and predicate letters are interpreted. There is no restriction on the sorts of objects that may be included, so long as the universe is not empty. In addition, a model supplies an interpretation (or semantic value) for the nonlogical symbols occurring in a given set of wffs, depending on the type of symbol.

Symbol	*Interpretation*
name letter	an individual object (e.g., the moon)
zero-place predicate letter	a truth value (T or F)
one-place predicate letter	a class of objects (e.g., the class of peple)
n-place predicate letter (for $n > 1$)	a relation between n objects (e.g., the relation asserting that a first object is greater than a second)

As with English names and predicates, distinct name letters can designate the same object and distinct predicate letters can stand for the same class or relation (for example, the predicates "person" and "human being"). However, if the same predicate letter represents predicates with a different number of places (for example, *Fa & Fbc*), then we assume the model interprets the letter in a different way for each distinct use. In general, we formalize different predicates using distinct letters or the same letter with distinct subscripts.

Given a model, every atomic formula φ is assigned a truth value according to the following rules:

> Rule 1. If φ consists of a single sentence letter, then its truth value is the one specified directly by the model.
>
> Rule 2. If φ consists of a predicate letter followed by a single name letter, then φ is assigned the value *T* if the object designated by the name letter is a member of the class designated by the predicate letter; otherwise φ is assigned the value *F*.
>
> Rule 3. If φ consists of a predicate letter followed by two or more name letters, then φ is assigned the value *T* if the objects designated by the name letters are in the relation designated by the predicate letter; otherwise φ is assigned the value *F*.

A statement assigned the value *T* in a model is said to be *true* in the model, and a statement assigned the value *F* in a model is said to be *false* in the model.

Example 6.8. Evaluate the following wffs in the given model:

(a) *Pc*
(b) *Pe*
(c) *Tec*
(d) *Tee*

Universe: every object in the world, including people and animals
c: Bill Clinton
e: the Empire State Building
P: the class of all people
T: the relation asserting that x is taller than y

(a) *T*, since Bill Clinton is a person.
(b) *F*, since the Empire State Building is not a person.
(c) *T*, since the Empire State Building is taller than Bill Clinton.
(d) *F*, since the Empire State Building is not taller than itself.

Once we can evaluate atomic formulas, the truth values of compound wffs *without* quantifiers are determined as in propositional logic using the conditions specified by the basic truth tables.

Example 6.9. Evaluate *(Pc & Tce)* ∨ *(~Pc & Tec)* in the model of Example 6.8.

> This wff is a disjunction. The first disjunct is a conjunction that is false in the model since its second conjunct *Tce* is false. The second disjunct is a conjunction that is false in the model since its first conjunct *~Pc* is the negation of a true atomic formula and is thus false. Since both disjuncts are false, the disjunction is false.

We now consider the truth conditions for quantified wffs; recall that this is the second difficulty mentioned at the beginning of this section. Suppose we have to evaluate the wff ∀ *x Sx* in the model of Example 6.8, where *S* is interpreted as the class of all saxophone players. Our wff asserts that "Everything is a saxophone player" and is false in the model; in particular, there exists an *instance* of the quantified wff ∀ *x Sx*, namely *Se*, that is false in the model. On the other hand, the wff ∃ *x Sx* is true in the model, since we can specify an instance of ∃ *x Sx*, namely *Sc*, that is true in the model.

Example 6.10. Evaluate ∀ *x (Px → Sx)* in the model:

> *Universe:* the class of all people, past and present
> *b*: George H.W. Bush
> *c*: Bill Clinton
> *P*: the class of all twentieth-century U.S. presidents
> *S*: the class of all saxophone players

> The instance *Pb → Sb* is false, since its antecedent is true (Bush has been a U.S. president in the twentieth-century), but its consequent is false (Bush is not a saxophone player). Since a universally quantified wff can be true only if every instance of it is true, the wff ∀*x (Px → Sx)* is false. Note that the wff is false even though one instance, *Pc → Sc*, is true.

In some cases, it may be necessary to consider a greater number of instances than the model initially allows. For example, when evaluating $\exists x \sim Px$ in the model of Example 6.10, it is not enough to just consider the false instances $\sim Pb$ and $\sim Pc$. In-deed, the fact that Bush and Clinton are twentieth-century U.S. presidents is not enough to deny the existence of *something* which is not a twentieth-century U.S. president; the quantifier ranges over all objects in the universe, including those not designated by a name letter. We overcome this difficulty by constructing a substitution instance, say $\sim Pa$. If a is in-terpreted as Walt Disney, then Pa is false, so $\sim Pa$ is true. An *a-variant* of a given model is the result of extending the model by interpreting a new name letter a. Since at least one a-variant assigns the value T to $\sim Pa$, the existentially quantified $\exists x \sim Px$ is evaluated as true in the model.

We may now state the truth conditions for quantified wffs more pre-cisely. Let M be any model and suppose α is a name letter. An α-variant of M is defined as any model that results from M by freely interpreting α as any object in the universe of M. If M did not assign an interpretation to α, then an α-variant is a slightly "richer" model than M. If M already assigned an interpretation to α, then an α-variant of M provides a new way of interpreting α, keeping everything else exactly as in M. Using the notion of an α-variant of M, the truth conditions for quantified sentences are defined as follows:

Rule 4. A universal quantification $\forall \beta \varphi$ is true in a model M if $\varphi(^\alpha/_\beta)$ is true in every α-variant of M, where α is the first name letter in the alphabet not occurring in φ and $\varphi(^\alpha/_\beta)$ is the result of replacing every occurrence of β in φ by α; if $\varphi(^\alpha/_\beta)$ is false in some α-variant of M, then $\forall \beta \varphi$ is false in M.

Rule 5. An existential quantification $\exists \beta \varphi$ is true in a model M if $\varphi(^\alpha/_\beta)$ is true in some α-variant of M, where α and $\varphi(^\beta/_\alpha)$ are as in Rule 4; if $\varphi(^\alpha/_\beta)$ is false in every α-variant of M, then $\exists \beta \varphi$ is false in M.

Note that α must be a name letter *not occurring* in φ; this restriction is crucial. For example, suppose φ is $\exists x Txc$ and we want to evaluate φ in the model of Example 6.8, where c is interpreted as Bill Clinton and T as the relation *taller than*. If α is the name letter c, the formula $\varphi(^\alpha/_\beta)$ be-comes Tcc, and since nothing is taller than itself, this formula is false in

every α-variant of the model. This results in an assignment of the value
F to the wff ∃ x Txc. However, this wff is true in the model, since there
are objects taller than Bill Clinton (for example, the Empire State Building).

Example 6.11. Evaluate ∃ x Gxa in the model:

> *Universe*: the class of all people
> a: Bill Clinton
> G: the relation that holds when x is younger than y

This is an existentially quantified wff and the first name letter not occurring in this wff is b. By Rule 5, this wff is true in M if there is at
least one b-variant of M in which Gba is true. Since there are many
people younger than Bill Clinton, there are many such b-variants and
∃ x Gxa is true in M. Note that, in this example, Bill Clinton is the
interpretation of the name letter a, not of the name letter c (which is
not interpreted by the model).

Example 6.12. Evaluate ∃ x Px & ∃ x Mx in the model:

> *Universe*: the class of all people, past and present
> M: the class of all male people
> P: the class of all twentieth-century U.S. presidents

This wff is a conjunction (not to be confused with the quantified wff
∃ x $(Px$ & $Mx))$ and is true if both conjuncts are true. The first conjunct is an existentially quantified wff ∃ x Px which contains no name
letters. So, we consider the values of Pa in the a-variants of M and,
since the class of twentieth-century U.S. presidents is nonempty,
there exist a-variants that interpret a as a twentieth-century U.S.
president. In such models Pa is true, and so ∃ x Px is true in M by
Rule 5. Similar reasoning shows that the second conjunct ∃ x Mx is
also true in M. Therefore, the wff ∃ x Px & ∃ x Mx is true as well.

Example 6.13. Evaluate ∀ x $(Wx \rightarrow Bx)$ in the model:

> *Universe*: the class of all living creatures
> B: the class of all blue things
> W: the class of all winged horses

This wff is true in M if the conditional $Wa \to Ba$ is true in every a-variant of M. No matter how we interpret a, the antecedent is false since there are no winged horses. Therefore, $Wa \to Ba$ is true in every a-variant of M and $\forall x\, (Wx \to Bx)$ is true in M.

The next example shows that Rules 4 and 5 can be applied more than once to evaluate wffs with nested quantifiers.

Example 6.14. Evaluate $\forall x\, (Ex \to \forall y\, Gxy)$ in the model:

Universe: the class of all positive integers
E: the class of all even integers
G: the relation that holds when x is greater than y

This wff is true in M if the conditional $Ea \to \forall y\, Gay$ is true in every a-variant of M. Let M^* be such a model. If a is interpreted as an odd integer, the antecedent Ea is false in M^* and the conditional is true. Now suppose that M^* interprets a as an even integer. Then the antecedent is true in M^*, and so the conditional is true in M^* only if the consequent $\forall y\, Gay$ is also true in M^*. This universally quantified wff is true in M^* just when the instance Gab (obtained by replacing y with the first available name letter b) is true in every b-variant of M^*. But any b-variant of M^* in which b is interpreted as an integer greater than the integer designated by a will make Gab false. Thus, if a is interpreted by M^* as an even integer, the conditional $Ea \to \forall y\, Gay$ is false in M^*. Therefore, since M^* is an a-variant of M, $\forall x\, (Ex \to \forall y\, Gxy)$ is false in M.

Similarly, one can show that wffs for which mixed consecutive quantifiers are reordered have different truth conditions. For example, in the model of Example 6.13, $\exists x\, \forall y\, Gxy$ is false in M, while $\forall y\, \exists x\, Gxy$ is true in M.

Refutation Trees

Recall that an argument form is valid if and only if there is no possible situation in which its premises are true while its conclusion is false. For predicate logic, a "possible situation" is characterized as a model that interprets all the nonlogical symbols occurring in the argument. An argument form of predicate logic is valid if and only if there is no such model in which its premises are assigned the value true while its conclusion is assigned the value false. For example, since there is a model in which the wff $\forall\, y\, \exists\, x\, Gxy$ is true while the wff $\exists\, x\, \forall\, y\, Gxy$ is false, any argument of the form $\forall\, y\, \exists\, x\, Gxy \,\vdash\, \exists\, x\, \forall\, y\, Gxy$ is invalid. In contrast, the "opposite" argument is valid.

Example 6.15. Show the validity of the argument form:

$$\exists\, x\, \forall\, y\, Gxy \,\vdash\, \forall\, y\, \exists\, x\, Gxy$$

If M is any model in which the premise is true, then the universe of M contains some object (call it "The One") which bears the relation designated by G to every other object in the universe. If this is the case, then for every object in the universe there is at least one object (namely "The One") in relation G to it. Therefore, the conclusion is also true in model M.

This characterization of validity is applicable to every argument form that can be expressed in the language of predicate logic. However, while this characterization of validity is significant from a theoretical point of view, it is not very efficient in practical application. In fact, predicate logic is worse than propositional logic in this regard. The method of truth tables is inefficient for questions involving numerous sentence letters, but it is a reliable algorithmic method. In contrast, our test for the validity of an argument form of predicate logic gives no finite bound on the number of models that must be considered. In particular, there is no limit to the size of a model's universe, and so no limit to the number of variants to be considered when evaluating a quantified wff. In addition, there is no limit to the possible interpretations of names and predicates.

You Need to Know

Unlike propositional logic or the logic of categorical statements, there is not and cannot be an algorithmic procedure for detecting invalidity in every predicate logic argument; that is, predicate logic is *undecidable.*

Fortunately, there are some rule-governed procedures for testing validity which yield an answer after finitely many steps for many (though not all) predicate logic argument forms. We consider one such procedure: a generalization of the refutation trees. We first note that for some predicate logic trees, the rules of propositional logic suffice.

Example 6.16. Use the refutation tree rules for propositional logic to determine the validity of the argument form:

$$\forall x\, Fx \rightarrow \forall x\, Gx,\ \sim\forall x\, Gx \vdash \sim\forall x\, Fx$$

1.	✓	$\forall x\, Fx \rightarrow \forall x\, Gx$			
2.		$\sim\forall x\, Gx$			
3.		$\sim\sim\forall x\, Fx$			
4.	$\sim\forall x\, Fx$	$1\rightarrow$		$\forall x\, Gx$	$1\rightarrow$
5.	X	$3,4\sim$	X		$2,4\sim$

All paths close, so the argument form is valid. Only propositional logic rules are needed, because the form is a substitution instance of *modus tollens* (MT) which is valid by propositional logic.

Our generalized refutation tree technique incorporates all the rules of propositional logic, together with four new rules. In particular, predicate logic has two new logical symbols (namely $\forall$ and $\exists$) and we need a rule to deal with the negated and unnegated sentences in which each occurs. The first is the universal quantification rule:

> *Universal Quantification* ($\forall$): If a wff of the form $\forall \beta\, \varphi$ appears on an open path and if α is a name letter occurring in some wff on the path, write $\varphi(^{\alpha}/_{\beta})$ (the result of replacing all occurrences of β in φ by α) at the bottom of the path. If (and only if) no wff containing a name letter appears on the path, choose a name letter α and write $\varphi(^{\alpha}/_{\beta})$ at the bottom of the path. In either case, do not check $\forall \beta\, \varphi$.

Intuitively, this rule formally expresses the fact that whatever is true of everything must be true of any particular individual. Recall that in each step of a refutation tree, we break down a wff into simpler wff(s), all of which are true if the original wff is true. If the wff in question is a universally quantified wff ∀ β φ, *every* wff obtainable from φ by replacing each occurrence of the variable β by a name α would be true; for if φ is true of everything, then it must be true of the object denoted by α, whatever that may be. In principle, there is no limit to the number of names that may be considered. However, in a refutation tree we are just interested in seeing whether every path will eventually close, so we focus on those names already appearing on some open path. The second part of the rule applies in those cases when no name yet occurs in an open path; in such cases we have to introduce some name to start the process, so we choose some (any) name α and write φ($^{α}/_{β}$) at the end of the open paths.

Example 6.17. Use a refutation tree to determine the validity of the argument form: ∀ x (Fx → Gx), ∀ x Fx ⊢ Ga.

1.	∀ x (Fx → Gx)	
2.	∀ x Fx	
3.	~ Ga	
4. ✓	Fa → Ga	1 ∀
5.	Fa	2 ∀

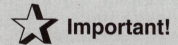

| 6. | ~ Fa | 4 → | Ga | 4 → |
| 7. | X | 5,6 ~ | X | 3,6 ~ |

Since the name letter *a* occurs in ~*Ga* in line 3, we use *a* to obtain lines 4 and 5 via universal quantification; in line 4, φ is *(Fx → Gx)*, α is *a*, β is *x*, and φ($^{α}/_{β}$) is *(Fa → Ga)*. The tree closes after applying the conditional rule at line 6, and so this argument form is valid.

★ Important!

When using the universal quantification rule, we *do not* check ∀ β φ, since no matter how many wffs we infer from it by ∀, we still have not exhausted all of its implications.

Example 6.18. Use a refutation tree to determine the validity of the argument form: Fa → Gb, ∀ x ~Fx ⊢ ~Gb.

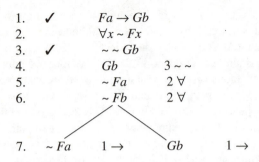

Applying nonbranching rules first, we check line 3 and obtain *Gb* in line 4 by double negation, and we apply universal quantification twice to derive ~ *Fa* and ~ *Fb* from line 2 in lines 5 and 6. Now, ∀ can be applied no more, since we have used it with every name letter occurring in wffs on the path; ∀ can introduce a name letter only if no name letters appear in wffs on the path. Thus, all that remains is to apply the conditional rule at line 7. This fails to close the tree, and so the form is invalid.

Every open path of a finished refutation tree can be interpreted as a model whose universe contains the objects mentioned by name on the path. The atomic wffs or negations of atomic wffs on the path indicate what is true of the objects in this model. In Example 6.18, the two open paths represent a model whose universe contains two objects *a* and *b*, where *b* is *G* and both *a* and *b* are not *F*. The wff *Fa* is false in this model and so the premise *Fa* → *Gb* is true. In addition, since the only objects in this universe are *a* and *b* and neither is *F*, the premise ∀ *x* ~*Fx* is true. Finally, the conclusion ~*Gb* is false since *b* is *G*. Thus, this model shows that at least one instance of this argument form has true premises and a false conclusion.

The next two rules are for negations of quantified wffs. The first expresses the equivalence between a negated existential quantification and a universally quantified negation. For example, "Nothing is a frog" can be formalized either as ~ ∃ *x Fx* or as ∀ *x* ~ *Fx*. Similarly, the second rule expresses the equivalence between a negated universal quantification and an existentially quantified negation.

Negated Existential Quantification (~ ∃): If an unchecked wff of form ~ ∃ β𝜑 appears on an open path, check it and write ∀ β ~ 𝜑 at the end of every open path containing the newly checked wff.

Negated Universal Quantification ($\sim \forall$): If an unchecked wff of form $\sim \forall \beta \varphi$ appears on an open path, check it and write $\exists \beta \sim \varphi$ at the end of every open path containing the newly checked wff.

Example 6.19. Use a refutation tree to show the validity of the argument form: $\forall x\, Fx \rightarrow \forall x\, Gx,\ \sim\exists x\, Gx \vdash \exists x\, \sim Fx$.

1.	✓	$\forall x\, Fx \rightarrow \forall x\, Gx$		
2.	✓	$\sim \exists x\, Gx$		
3.		$\sim\exists x \sim Fx$		
4.		$\forall x \sim Gx$	$2 \sim \exists$	

5.	✓	$\sim \forall x\, Fx$	$1\rightarrow$	$\forall x\, Gx$	$1 \rightarrow$
6.		$\exists x \sim Fx$	$5 \sim \forall$	Ga	$5\, \forall$
7.		X	$3, 6 \sim$	$\sim Ga$	$4\, \forall$
8.				X	$6, 7 \sim$

The final quantifier rule allows us to simplify a wff by dropping an initial existential quantifier.

Existential Quantification ($\exists$): If an unchecked wff of the form $\exists \beta \varphi$ appears on an open path, check it. Then choose a name letter that does not appear anywhere on that path and write $\varphi(^\alpha/_\beta)$ (the result of replacing every occurrence of β in φ by α) at the bottom of every open path that contains the newly checked wff.

The intuitive rationale for this rule is that if an existentially quantified wff $\exists x\, \varphi$ is true, then there must exist some object of which φ is true. We may not know what that object is, but we name it. Since we do not want to make any specific assumptions about the object, we choose a name letter not yet appearing on the path containing $\exists x\, \varphi$.

Remember

When using the existential quantification rule, we must chose a *new*
name letter.

Example 6.20. Use a refutation tree to determine the validity of the argument form: ∃ x Fx, ∃ x Gx ⊢ ∃ x (Fx & Gx).

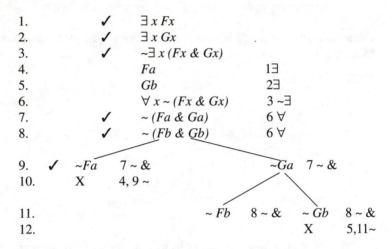

1.   ✓   ∃ x Fx
2.   ✓   ∃ x Gx
3.   ✓   ~∃ x (Fx & Gx)
4.   Fa   1∃
5.   Gb   2∃
6.   ∀ x ~ (Fx & Gx)   3 ~∃
7.   ✓   ~ (Fa & Ga)   6 ∀
8.   ✓   ~ (Fb & Gb)   6 ∀

9.   ✓   ~Fa   7 ~ &   ~Ga   7 ~ &
10.   X   4, 9 ~

11.   ~ Fb   8 ~ &   ~ Gb   8 ~ &
12.   X   5,11~

The form is invalid. In fact, the open path represents a universe containing two objects, *a* and *b*, such that *a* is *F* but not *G* and *b* is *G* but not *F*. Notice that in applying existential quantification, we introduce a new name letter *a* at line 4 and a *different* new name letter *b* at line 5. Also the tree is complete, since no further rules apply to the open path.

The refutation tree test for the validity of an argument form without premises in predicate logic is the same as for propositional logic: negate the conclusion and apply the rules to construct a tree. The form is valid if and only if all paths of the finished tree are closed.

Example 6.21. Use a refutation tree to show the validity of the argument form: ⊢ ~ (∃ x Fx & ∀ x ~ Fx).

1.   ✓   ~ ~ (∃ x Fx & ∀ x ~ Fx)
2.   ✓   ∃ x Fx & ∀ x ~ Fx   1 ~ ~
3.   ✓   ∃ x Fx   2 &
4.   ∀ x ~ Fx   2 &
5.   Fa   3 ∃
6.   ~ Fa   4 ∀
7.   X   5, 6 ~

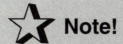

Note!

The quantification rules work the same way for wffs containing multiple quantifiers as they do for wffs containing single quantifiers.

Example 6.22. Use a refutation tree to show the validity of the argument form: $\exists x \forall y\, Lxy \vdash \forall x \exists y\, Lyx$.

1.	✓	$\exists x \forall y\, Lxy$	
2.	✓	$\sim\forall x \exists y\, Lyx$	
3.		$\forall y\, Lay$	1 $\exists$
4.	✓	$\exists x \sim \exists y\, Lyx$	2 $\sim\forall$
5.	✓	$\sim \exists y\, Lyb$	4 $\exists$
6.		$\forall y \sim Lyb$	5 $\sim \exists$
7.		$\sim Lab$	6$\forall$
8.		Lab	3$\forall$
9.		X	7, 8 $\sim$

Example 6.23. Try to use a refutation tree to show the invalidity of the argument form: $\forall x \exists y\, Lxy \vdash Laa$.

1.		$\forall x \exists y\, Lxy$	
2.		$\sim Laa$	
3.	✓	$\exists y\, Lay$	1 $\forall$
4.		Lab	3 $\exists$
5.	✓	$\exists y\, Lby$	1 $\forall$
6.		Lbc	5 $\exists$
7.	✓	$\exists y\, Lcy$	6 $\forall$
8.		Lcd	7 $\exists$
		$\vdots$	

The tree never ends and is infinitely long. We apply the universal quantification rule to line 1 and generate a new existentially quantified wff at line 3. Applying the existential quantification rule to line 3 introduces a new name letter b at line 4. This cycle of applying first universal and then existential quantification continues forever, so the path never closes and the tree never returns an answer about the validity of this argument form. Intuitively, this argument form is invalid; if we interpret L as "is the father of", we see that from the assumption that everything has a father, it does not follow that a is his own father.

This last example illustrates the *undecidability* of predicate logic. In particular, it is impossible to produce a finite algorithm that correctly returns the answer "valid" or "invalid" for every argument form. Notice, however, that the tree in Example 6.23 does not give the wrong answer, rather it does not give any. In fact, the answers given by the refutation tree test are always correct.

Identity

Special symbols may be added to the language of predicate logic for special purposes. Perhaps the most useful is the identity predicate, "=", which means "is identical to" or "is the same thing as". In logic, the name letters a and b in the wff $a = b$ may be broadly interpreted as any sort of object. We might interpret a, for example, as "Mark Twain" and b as "Samuel Clemens", to that $a = b$ means "Mark Twain is identical to (i.e., the same person as) Samuel Clemens".

You Need to Know

The identity predicate is special because, like the logical symbols and unlike ordinary predicate letters, its interpretation is fixed and *always* means "is identical to". Also, unlike other two-place predicates, "=" is not written before, but *between* the two name letters to which it applies.

Example 6.24. Interpreting the letters c, h, and t as "Samuel Clemens", "*Huckleberry Finn*" (the book) and "Mark Twain", and W as the two-place predicate "wrote", formalize the following sentences.

(a) Mark Twain is not Samuel Clemens.
(b) Mark Twain exists.
(c) Only Mark Twain wrote *Huckleberry Finn*.
(d) At least two things exist.
(e) At most one thing exists.
(f) Exactly one thing exists.
(g) Exactly two things exist.

(a) $\sim t = c$. This is often written $t \neq c$. In accordance with the formation rules, there are no parentheses around the atomic wff $t = c$; the $\sim$ applies to the entire wff $t = c$.
(b) $\exists x \, (x = t)$
(c) $Wth \,\&\, \forall x \, (Wxh \rightarrow x = t)$
(d) $\exists x \, \exists y \sim (x = y)$.
(e) $\forall x \, \forall y \, (x = y)$. This says, "For any objects x and y, x is identical to y", and so the universe contains at most one thing.
(f) $\exists x \, \forall y \, (x = y)$. This says, "There exists an object to which all objects are identical".
(g) $\exists x \, \exists y \, (\sim x = y \,\&\, \forall z \, (z = x \vee z = y))$. The first clause "$\sim x = y$" asserts that x and y are two different things, while the second clause asserts that there is nothing other than x or y.

Semantically, "$=$" is interpreted in every model as the identity relation. Thus, every object in the domain bears that relation to itself and to nothing else. The refutation tree technique is extended to the identity predicate under this intended interpretation with the new rules:

Identity ($=$): If a wff of the form $\alpha = \beta$ appears on an open path and another wff φ containing either α or β appears unchecked on that path, write at the bottom of the path any wff not already on the path which results from replacing one or more occurrences of these name letters by the other in φ. Do not check either $\alpha = \beta$ or φ.

Negated Identity ($\sim =$): Close any open path on which a wff of the form $\sim \alpha = \alpha$ occurs.

The identity rule expresses the idea that if an identity holds between the objects designated by two names α and β, then those names are interchangeable in every wff φ. The negated identity rule expresses the idea that every object is self-identical.

Example 6.25. Use a refutation tree to determine the validity of the argument form: $\sim a = b$, $\sim b = c \vdash \sim a = c$.

1.		$\sim a = b$	
2.		$\sim b = c$	
3.	✓	$\sim \sim a = c$	
4.		$a = c$	$3 \sim \sim$
5.		$\sim c = b$	$1, 4 =$
6.		$\sim b = a$	$2, 4 =$

We have applied the identity rule for line 4 to every wff to which it can be applied, but still the tree is not closed. The form is therefore invalid. The tree shows us that the premises may be true and the conclusion false in a model containing objects a, b, c such that a and c are identical, but distinct from b.

Example 6.26. Use a refutation tree to show the validity of the argument form: $\vdash \forall x \, \forall y \, (x = y \rightarrow y = x)$.

1.	✓	$\sim \forall x \, \forall y \, (x = y \rightarrow y = x)$	
2.		$\exists x \sim \forall y \, (x = y \rightarrow y = x)$	$1 \sim \forall$
3.	✓	$\sim \forall y \, (a = y \rightarrow y = a)$	$2 \, \exists$
4.	✓	$\exists y \sim (a = y \rightarrow y = a)$	$3 \sim \forall$
5.	✓	$\sim (a = b \rightarrow b = a)$	$4 \, \exists$
6.		$a = b$	$5 \sim \rightarrow$
7.		$\sim b = a$	$5 \sim \rightarrow$
8.		$\sim a = a$	$6, 7 =$
9.		X	$8 \sim =$

Chapter 7
THE PREDICATE CALCULUS

IN THIS CHAPTER:

- ✔ *Reasoning in Predicate Logic*
- ✔ *Inference Rules for the Universal Quantifier*
- ✔ *Inference Rules for the Existential Quantifier*
- ✔ *Theorems and Quantifier Equivalence Rules*
- ✔ *Inference Rules for the Identity Predicate*

Reasoning in Predicate Logic

We extend the propositional calculus to predicate logic with identity. As in the propositional calculus, the characterization of validity via truth values or refutation trees is in principle adequate. However, they do not capture the inferential processes of typical arguments, in which a conclusion is established by *deducing* it from some premises.

The *predicate calculus* includes all the basic introduction and elimination rules of the propositional calculus, and so all the derived rules. In addition, the predicate calculus has new introduction and elimination rules for the quantifiers and for the identity predicate.

This set of rules is both *valid* (in the sense that the rules generate only argument forms that are valid by virtue of the semantics of the quantifiers, the truth-functional connectives, and the identity predicate) and *complete* (in the sense that they generate all valid argument forms). Validity and completeness are important for guaranteeing that the inferential techniques of the predicate calculus identify the same class of valid argument forms as the semantic techniques of Chapter 6. This also means that the predicate calculus is *undecidable*, since the property of being a valid predicate logic argument form is undecidable.

Inference Rules for the Universal Quantifier

We first consider a proof in the predicate calculus that uses only the rules of propositional logic. Despite their greater internal complexity, wffs of predicate logic are treated the same as the wffs of propositional logic by the rules of the propositional calculus.

Example 7.1. Prove: $\sim Fa \vee \exists x\, Fx, \exists x\, Fx \rightarrow P \vdash Fa \rightarrow P$

1. $\sim Fa \vee \exists x\, Fx$ A
2. $\exists x\, Fx \rightarrow P$ A
3. Fa H (for $\rightarrow$ I)
4. $\sim\sim Fa$ 3 DN
5. $\exists x\, Fx$ 1, 4 DS
6. P 2, 5 $\rightarrow$ E
7. $Fa \rightarrow P$ 3 – 6 $\rightarrow$ I

The first of our new rules of inference is the elimination rule for the universal quantifier. This rule formally expresses the fact that whatever is true of everything must be true of a particular individual.

Universal Elimination ($\forall$ E): From a universally quantified wff $\forall \beta\, \varphi$ we may infer any wff of the form $\varphi(^{\alpha}/_{\beta})$ which results from replacing every occurrence of the variable β in φ by some name letter α. Sometimes this rules is called *universal instantiation*.

Example 7.2. Prove: $\forall x (Fx \rightarrow Gx)$, $\forall x Fx \vdash Ga$

1. $\forall x (Fx \rightarrow Gx)$ A
2. $\forall x Fx$ A
3. $Fa \rightarrow Ga$ 1 $\forall$ E
4. Fa 2 $\forall$ E
5. Ga 3, 4 $\rightarrow$ E

Example 7.3. Prove: $\sim Fa \vdash \sim \forall x Fx$

1. $\sim Fa$ A
2. $\forall x Fx$ H (for $\sim$I)
3. Fa 2 $\forall$ E
4. $Fa \;\&\; \sim Fa$ 1, 3 & I
5. $\sim \forall x Fx$ 2 – 4 $\sim$ I

Since the conclusion is negated, we use *reductio ad absurdum*. We hypothesize the conclusion without its negation at line 2, leading to a contradiction at line 4.

We now consider the introduction rule for the universal quantifier, which is used to prove universally quantified conclusions. Intuitively, universal introduction asserts that if we can prove something about an individual a without using any distinguishing assumptions about *a*, then what we have proved for *a* holds for everything.

Example 7.4. Prove:

$\forall x (Fx \rightarrow Gx)$, $\forall x (Gx \rightarrow Hx) \vdash \forall x (Fx \rightarrow Hx)$.

1. $\forall x (Fx \rightarrow Gx)$ A
2. $\forall x (Gx \rightarrow Hx)$ A
3. $Fa \rightarrow Ga$ 1 $\forall$E
4. $Ga \rightarrow Ha$ 2 $\forall$ E
5. $Fa \rightarrow Ha$ 3, 4 HS
6. $\forall x (Fx \rightarrow Hx)$ 5 $\forall$ I

The name letter *a* introduced by $\forall$ E at lines 3 and 4 specifies an individual about whom we have made no assumptions (*a* does not oc-

cur in lines 1 and 2). Indeed, we could replace every *a* in the proof
with any other name letter, and so *a* functions in this proof as a representative of all individuals.

Formally, the universal introduction rule (also called *universal generalization*) is stated as follows:

Universal Introduction ($\forall$ I): From a wff ϕ containing a name letter
α not occurring in any assumptions or in any hypothesis in effect
at the line on which ϕ occurs, we may infer any wff of the form
$\forall \beta \phi(^{\beta}/_{\alpha})$, where $\phi(^{\beta}/_{\alpha})$ is the result of replacing all occurrences of
α in ϕ by some variable β not already in ϕ.

Requiring α to not occur in any assumption or in any hypothesis in
effect at the line on which ϕ occurs ensures that we assume nothing distinguishing α from any other individual. Universal introduction must be
applied strictly as stated; the following qualifications are important:

1. *The name letter α may not appear in any assumption.* For example, the following derivation is invalid:

   1. *Fa*   A
   2. $\forall x Fx$ 1 $\forall$ I (incorrect)

2. *The name letter α may not appear in any hypothesis in effect at
   the line on which ϕ occurs.* Recall that a hypothesis is *in effect* if
   the vertical line beginning with the hypothesis extends down
   through the line; for example, the following is invalid:

   1. $\forall x (Fx \to Gx)$ A
   2. $Fa \to Ga$ 1 $\forall$ E
   3. | *Fa*   H (for $\to$ I)
   4. | *Ga*   2, 3 $\to$ E
   5. | $\forall x Gx$ 4 $\forall$ I (incorrect)
   6. $Fa \to \forall x Gx$ 3 – 5 $\to$ I

3. $\phi(^{\beta}/_{\alpha})$ *is the result of replacing **all** occurrences of α in ϕ by some
   variable β.* For example, the following is invalid:

1. $\forall x\ Lxx$ A
2. Laa $1\ \forall$ E
3. $\forall z\ Lza$ $2\ \forall$ I (incorrect)

The following two examples illustrate correct uses of the universal quantifier rules.

Example 7.5. Prove: $\forall x\ (Fx\ \&\ Gx) \vdash \forall x\ Fx\ \&\ \forall x\ Gx$

1. $\forall x\ (Fx\ \&\ Gx)$ A
2. $Fa\ \&\ Ga$ $1\ \forall$ E
3. Fa 2 &E
4. Ga 2 &E
5. $\forall x\ Fx$ $3\ \forall$ I
6. $\forall x\ Gx$ $4\ \forall$ I
7. $\forall x\ Fx\ \&\ \forall x\ Gx$ 5, 6 &I

Using a as a representative, we instantiate $\forall x\ (Fx\ \&\ Gx)$ at line 2. We make no assumptions or hypotheses about a so the applications of $\forall$ I at lines 5 and 6 are legitimate.

Example 7.6. Prove:

$$\forall x\ (Fx \rightarrow (Gx \lor Hx)), \forall x{\sim} Gx \vdash \forall x\ Fx \rightarrow \forall x\ Hx$$

1. $\quad \forall x\ (Fx \rightarrow (Gx \lor Hx))$ A
2. $\quad \forall x{\sim} Gx$ A
3. $\quad Fa \rightarrow (Ga \lor Ha)$ $1\ \forall$ E
4. $\quad {\sim} Ga$ $2\ \forall$ E
5. $\quad\quad \forall x\ Fx$ H (for $\rightarrow$ I)
6. $\quad\quad Fa$ $5\ \forall$ E
7. $\quad\quad Ga \lor Ha$ $3,6 \rightarrow$ E
8. $\quad\quad Ha$ 4,7 DS
9. $\quad\quad \forall x\ Hx$ $8\ \forall$ I
10. $\forall x\ Fx \rightarrow \forall x\ Hx$ $5\text{-}9, \rightarrow$ I

We instantiate the assumptions at line 2 using the new name letter a. Since the conclusion is conditional, we hypothesize its antecedent at line 5 in order to derive its consequent and then apply conditional introduction.

Inference Rules for the Existential Quantifier

The existential introduction rule captures the intuitive idea that from a premise that an individual has a certain property, it follows that *something* has that property. Formally, we have:

> *Existential Introduction* (∃ I): Given a wff φ containing some name letter α, we may infer any wff of the form ∃ β φ($^β/_α$), where φ($^β/_α$) is the result of replacing one or more occurrences of α in φ by some variable β not already in φ.

There are several important things to note about this rule:

1. *Unlike* ∀ I, ∃ I *places no restrictions on previous occurrences of the name letter α.*

2. *Unlike* ∀ I, *the variable β need not replace every occurrence of α in φ.* For example, the following proof is correct:

   1. *Fa & Ga*   A
   2. *∃ x (Fx & Ga)*   1 ∃ I

3. *Like* ∀ I, ∃ I *allows us to introduce only one quantifier at a time, and only at the leftmost position in a formula.* For example, the following inference is incorrect:

   1. *Fa → Ga*   A
   2. *∃ x Fx → Ga*   1 ∃ I (incorrect)

 **Note!**

Existential Introduction highlights two important presuppositions implicit in the semantics of predicate logic:
   1. All proper names refer to existing individuals.
   2. At least one individual exists.

Example 7.7. Prove: $\sim \exists\, x\, Fx \vdash \forall\, x \sim Fx$

1.	$\sim \exists\, x\, Fx$	A
2.	Fa	H (for $\sim$ I)
3.	$\exists\, x\, Fx$	2 $\exists$ I
4.	$\exists\, x\, Fx \,\&\, \sim \exists\, x\, Fx$	1,3 &I
5.	$\sim Fa$	2 – 4, $\sim$ I
6.	$\forall\, x \sim Fx$	5 $\forall$ I

We obtain the universally quantified conclusion $\forall\, x \sim Fx$ by deducing $\sim Fa$ and then applying universal introduction. Since $\sim Fa$ is negated, the $\sim$ I strategy is appropriate and we hypothesize Fa at line 2. Although our hypothesis contains the name letter a, it is no longer in effect at line 5 and a does not occur in the assumptions, so applying universal introduction is legitimate.

We need existential elimination to reason from existential premises and this rule uses hypothetical reasoning. We first choose an individual to represent one of the things having the property specified by the premise and hypothesize that this individual in fact has the property. Without making any additional assumptions about this individual, we derive the conclusion we are trying to prove. Since we have only assumed the property that the existential premise ascribes to *something*, deriving the desired conclusion from our hypothesis shows that no matter which individual has the property, our conclusion must be true. We are therefore entitled to discharge the hypothesis and assert our conclusion based on the existential premise alone. Formally, we have:

Existential Elimination ($\exists$ E): Given an existentially quantified wff $\exists\, \beta\, \varphi$ and a derivation of some conclusion ψ from a hypothesis of the form $\varphi(^{\alpha}/_{\beta})$ (the result of replacing every occurrence of the variable β in φ by some name letter α not already in φ), we may discharge $\varphi(^{\alpha}/_{\beta})$ and reassert ψ. *Restriction:* The name letter α may not occur in ψ, nor in any assumption, nor in any hypothesis that is in effect at the line at which $\exists$ E is applied.

We call $\varphi(^{\alpha}/_{\beta})$ a *representative instance* of $\exists\, \beta\, \varphi$, since α represents one of that things with the property specified by φ. There are several things to note about this rule:

1. *The name letter α must not already occur in φ.* This is to prevent mistakes such as the following:

1. ∀ x ∃ y Fyx	A
2. ∃ y Fya	1 ∀ E
3. | *Faa*	H (for ∃ E)
4. | ∃ x Fxx	3 ∃ I
5. ∃ x Fxx	2, 3 – 4, ∃ E (incorrect)

Intuitively, assuming everything has a father (∀ x ∃ y Fyx) does not imply that something is its own father (∃ x Fxx). Since *a* already occurs in ∃ y Fya using *a* at line 3 to represent the individual(s) designated by *y* in this wff deprives the hypothetical derivation (lines 3 and 4) of generality, invalidating the use of ∃ E at line 5.

2. *The name letter α must not occur in ψ (the conclusion of the hypothetical derivation).* If this restriction is violated, the following sort of mistake may occur:

1. ∃ x Hxx	A
2. | *Haa*	H (for ∃ E)
3. | ∃ x Hax	3 ∃ I
4. ∃ x Hax	1, 2 – 3 ∃ E (incorrect)

Intuitively, assuming something hit itself (∃ x Hxx) does not imply that Alice hit something (∃ x Hax). Since *a* occurs in ∃ x Hax, using *a* at line 2 deprives the hypothetical derivation of generality, invalidating the use of ∃ E at line 4.

3. *α must not occur in any assumption.* For example, the following derivation is incorrect:

1. ∃ x Gx	A
2. *Fa*	A
3. | *Ga*	H (for ∃ E)
4. | *Fa* & *Ga*	2,3 & I
5. | ∃ x (Fx & Gx)	4 ∃ E
6. ∃x (Fx & Gx)	1, 3 – 5 ∃ E (incorrect)

Intuitively, assuming something is a giraffe ($\exists x\ Gx$) and Amos is a frog (Fa) does not imply that something is both a frog and a giraffe $\exists x\ (\ Fx\ \&\ Gx\ )$. Since a occurs in the assumption Fa, using a at line 3 deprives the hypothetical derivation of generality, invalidating the use of $\exists$ E at line 6.

4. α *must not occur in any hypothesis in effect at the line at which* $\exists$ E *is applied.* This is essentially the same provision as condition 3 only for hypotheses rather than assumptions.

The following proofs show the correct use of existential elimination.

Example 7.8. Prove: $\forall x\ (Fx \rightarrow Gx),\ \exists x\ Fx \vdash \exists x\ Gx$

1.	$\forall x\ (Fx \rightarrow Gx)$	A
2.	$\exists x\ Fx$	A
3.	$\quad Fa$	H (for $\exists$ E)
4.	$\quad Fa \rightarrow Ga$	1 $\forall$ E
5.	$\quad Ga$	3, 4 $\rightarrow$ E
6.	$\quad \exists x\ Gx$	5 $\exists$ I
7.	$\exists x\ Gx$	2, 3 – 6 $\exists$ E

We must perform existential introduction ($\exists$ I) at line 6 *before* applying existential elimination ($\exists$ E), otherwise the conclusion of our hypothetical derivation would be Ga, which contains a, and existential elimination would not apply.

Example 7.9. Prove: $\forall x\ (Fx \rightarrow \sim Gx) \vdash \sim \exists x\ (Fx\ \&\ Gx)$

1.	$\forall x\ (Fx \rightarrow \sim Gx)$	A
2.	$\quad \exists x\ (Fx\ \&\ Gx)$	H (for $\sim$ I)
3.	$\quad\quad Fa\ \&\ Ga$	H (for $\exists$ E)
4.	$\quad\quad Fa \rightarrow \sim Ga$	1 $\forall$ E
5.	$\quad\quad Fa$	3 & E
6.	$\quad\quad \sim Ga$	4, 5 $\rightarrow$ E
7.	$\quad\quad Ga$	3 & E
8.	$\quad\quad Ga\ \&\ \sim Ga$	6, 7 CON
9.	$\quad Ga\ \&\ \sim Ga$	2, 3 – 8 $\exists$ E
10.	$\sim \exists x\ (Fx\ \&\ Gx)$	2 – 9 $\sim$ I

Keep in mind the following when constructing proofs:

1. *All four quantifier rules only operate on a quantifier at the left-most position of a formula.*

2. *To prove a universally or existentially quantified conclusion, the typical strategy is first to prove a formula from which this conclusion can be obtained by universal introduction* ($\forall$ I) *or existential introduction* ($\exists$ I). We work with the following:

To Prove:	First Prove:
$\exists x\,Fx$	Fa
$\forall x\,(Fx \rightarrow Gx)$	$Fa \rightarrow Ga$
$\forall x \sim Fx$	$\sim Fa$
$\forall x\,y\,Fxy$	$\exists y\,Fay$
$\exists y\,Fxy$	Fab
$\exists x\,Fxx$	Faa

3. *If the conclusion is in the form of a negation, conjunction, disjunction, conditional, or biconditional, then employ the propositional calculus strategies for proving the conclusion.*

Theorems and Quantifier Equivalence Rules

As in the propositional calculus, we prove *theorems* of the predicate calculus without making assumptions. These theorems are exactly the *logical truths* of predicate logic; i.e., those wffs that are true in every model regardless of the meaning assigned their nonlogical symbols. All theorems of the propositional calculus are also theorems of the predicate calculus, but the predicate calculus has additional theorems.

Example 7.10. Prove the theorem: $\vdash \forall x\,(Fx \rightarrow Fx)$

1.	Fa	H (for $\rightarrow$ I)
2.	$Fa \rightarrow Fa$	$1 - 1 \rightarrow$ I
3.	$\forall x\,(Fx \rightarrow Fx)$	$2\,\forall$ I

Note that the final use of universal introduction is legitimate; although *a* occurs in the hypothesis *Fa*, the hypothesis is discharged at line 2.

Example 7.11. Prove the theorem: $\vdash \forall x\, Fx \rightarrow Fa$

1.	$\forall x\, Fx$	H (for $\rightarrow$ I)
2.	Fa	1 $\forall$ E
3.	$\forall x\, Fx \rightarrow Fa$	1 – 2 $\rightarrow$ I

Since the theorem is a conditional statement, we use the conditional proof strategy by hypothesizing the antecedent, deriving the consequent, and applying conditional introduction.

Example 7.12. Prove the theorem: $\vdash \sim (\forall x\, Fx\, \&\, \exists x \sim Fx)$

1.	$\forall x\, Fx\, \&\, \exists x \sim Fx$	H (for $\sim$ I)
2.	$\forall x\, Fx$	1 & E
3.	$\exists x \sim Fx$	1 & E
4.	$\sim Fa$	H (for $\exists$ E)
5.	Fa	2 $\forall$ E
6.	$P\, \&\, \sim P$	4, 5 CON
7.	$P\, \&\, \sim P$	3, 4 – 6 $\exists$ E
8.	$\sim(\forall x\, Fx\, \&\, \exists x \sim Fx)$	1 – 7 $\sim$ I

The following four equivalences express important relationships between the universal and existential quantifiers. These correspond to the refutation rules for negated quantifiers given in Chapter 6 and the proof of the first of these is given in Example 7.13.

$\vdash \sim \forall x \sim Fx \leftrightarrow \exists x\, Fx$ and $\vdash \forall x\, Fx \leftrightarrow \sim\exists x \sim Fx$

$\vdash \sim\forall x\, Fx \leftrightarrow \exists x \sim Fx$ and $\vdash \forall x \sim Fx \leftrightarrow \sim\exists x\, Fx$

Example 7.13. Prove the equivalence: $\vdash \sim \forall x \sim Fx \leftrightarrow \exists x\, Fx$

1.	$\sim \forall x \sim Fx$	H (for $\rightarrow$ I)
2.	$\sim\exists x\, Fx$	H (for $\sim$ I)
3.	Fa	H (for $\sim$ I)
4.	$\exists x\, Fx$	3 $\exists$ I
5.	$\exists x\, Fx\, \&\, \sim \exists x\, Fx$	2, 4 & I
6.	$\sim Fa$	3 – 5 $\sim$ I
7.	$\forall x \sim Fx$	6 $\forall$ I
8.	$\forall x \sim Fx\, \&\, \sim \forall x \sim Fx$	1, 7 & I
9.	$\sim \sim\exists x\, Fx$	2 – 8 $\sim$ I
10.	$\exists x\, Fx$	9 $\sim$ E

11. ~∀ x ~ Fx → ∃ x Fx	1 – 10 → I
12. ∃ x Fx	H (for → I)
13. Fa	H (for ∃ E)
14. ∀ x ~ Fx	H (for ~ I)
15. ~ Fa	14 ∀ E
16. Fa & ~ Fa	13, 15 & I
17. ~ ∀ x ~ Fx	14 – 16 ~I
18. ~ ∀ x ~ Fx	12, 13 – 17 ∃ E
19. ∃ x Fx → ~ ∀ x ~ Fx	12 – 18 → I
20. ~ ∀ x ~ Fx ↔ ∃ x Fx	11, 19 ↔ I

The four quantifier equivalences are the basis for the most important derived rules of the predicate calculus. Before stating the rules, we introduce some terminology. A formula that results from a wff by removing an initial quantifier-plus-variable is called an *open formula on the variable*. For example, removing ∃ x from ∃ x(Fx & Gx) results in the open formula *(Fx & Gx)* on x. Note that an open formula is not itself a wff, since it contains unquantified variables.

Now, in the proof given in Example 7.13, all occurrences of the formula *Fx* could have been replaced by any other open formula on x, and the result would still be a valid proof. Similarly, the variable x could have been uniformly replaced in the proof by any other variable without affecting the validity of the proof. Thus, Example 7.13 shows that any wff of the form ~∀ β ~ φ, where β is a variable and φ is an open formula on that variable, is logically equivalent to ∃ β φ. The other quantifier equivalences have the same generality, yielding the following derived rule:

Quantifier Exchange (QE): If φ is an open formula on a variable β and if one of the following wffs is a subwff of some wff ψ, we may validly infer from ψ the result of replacing one or more occurrences of this wff by the other member of its pair:

∀ β φ	and	~ ∃ β ~ φ
∀ β ~ φ	and	~ ∃ β φ
~∀ β φ	and	∃ β ~ φ
~∀ β ~ φ	and	∃ β φ

As with the derived rules of the propositional calculus (see Chapter 4), quantifier exchange (QE) enables us to shorten and simplify proofs. However, anything provable by QE is already provable by the four quantifier rules, together with the ten rules of propositional calculus.

Example 7.14. Prove: $\forall x (Fx \rightarrow \sim Gx) \vdash \sim \exists x (Fx \& Gx)$

We are reproving the validity of this argument form, which was first considered in Example 7.9.

1. $\forall x (Fx \rightarrow \sim Gx)$	A
2. $Fa \rightarrow \sim Ga$	1 $\forall$ E
3. $\sim(Fa \& \sim \sim Ga)$	2 theorem of prop. calculus
4. $\sim(Fa \& Ga)$	3 DN
5. $\forall x \sim(Fx \& Gx)$	4 $\forall$ I
6. $\sim \exists x (Fx \& Gx)$	5 QE

Inference Rules for the Identity Predicate

We introduce the introduction and elimination rule for the identity predicate, both of which are related in an obvious way to the identity rules for the refutation tree method.

> *Identity Introduction* (=I): For any name letter α, we may assert $\alpha = \alpha$ at any line of a proof.
> *Identity Elimination* (=E): If φ is a wff containing a name letter α, then from φ and either $\alpha = \beta$ or $\beta = \alpha$ we may infer $\varphi(\beta/_\alpha)$, where $\varphi(\beta/_\alpha)$ is the result of replacing one or more occurrences of α in φ by β.

The identity elimination rule is also called *substitutivity of identity* and formally expresses the familiar idea that if $a = b$, then the names a and b are interchangeable.

Example 7.15. Prove the theorem: $\vdash \forall x \, x = x$

1. $a = a$	= I
2. $\forall x \, x = x$	1 $\forall$ I

Example 7.16. Prove: $Fa, a = b \vdash Fb$

1. Fa	A
2. $a = b$	A
3. Fb	1, 2 = E

Example 7.17. Prove: $Fa, {\sim}Fb \vdash {\sim}a = b$

1.	Fa	A
2.	${\sim}Fb$	A
3.	$a = b$	H (for ~I)
4.	Fb	1, 3 =E
5.	$Fb \And {\sim}Fb$	2, 4 &I
6.	${\sim}a = b$	3 − 5 ~I

Example 7.18. Prove the theorem: $\vdash \forall x \, \forall y \, (x = y \rightarrow y = x)$

1.	$a = b$	H (for $\rightarrow$ I)
2.	$a = a$	=I
3.	$b = a$	1, 2 = E
4.	$a = b \rightarrow b = a$	1 − 3, $\rightarrow$ I
5.	$\forall y \, (a = y \rightarrow y = a)$	5 $\forall$ I
6.	$\forall x \, \forall y \, (x = y \rightarrow y = x)$	6 $\forall$ I

Chapter 8
FALLACIES

Classification of Fallacies

In this and the following chapter, we consider various techniques for evaluating nondeductive arguments. We begin with the study of fallacies. The study is informal in style, but augments our intuition with a general account of the most common mistakes of ordinary reasoning.

Fallacies (from the Latin verb *fallere* meaning "to deceive") are mistakes in arguments that affect their cogency. Fallacious arguments are often deceptive because they superficially appear to be good arguments, but deception is not a necessary condition of a fallacy. Whenever we reason invalidly or irrelevantly, accept premises we should not, or fail to use available facts, we commit a fallacy.

We study six classes of fallacies:

1. *Fallacies of relevance* occur when the premises of an argument have no bearing upon its conclusion and may have a distractive element diverting attention away from this problem.
2. *Circular reasoning* is the fallacy of assuming what we are trying to prove.
3. *Semantic fallacies* result when the language employed to construct arguments has multiple meanings or is excessively vague in a way that interferes with assessment of the argument.
4. *Inductive fallacies* occur when the probability of an argument's conclusion, given its premises, is low, or at least less than the arguer supposes.
5. *Formal fallacies* occur when we misapply a valid rule of inference or follow a rule which is invalid.
6. Finally, there is a class of mistakes classified as fallacies consisting of arguments with *false premises.*

In this chapter we provide a representative, but not exhaustive, survey of these fallacies. Many logic texts employ Latin expressions to label fallacies; we furnish Latin terms when such usage is customary.

Fallacies of Relevance

Fallacies of relevance occur when the premises of an argument have no bearing upon its conclusion and are often called *non sequiturs* (meaning "it does not follow"). We distinguish a number of fallacies of relevance, but the generic form is the same in each case.

Ad hominem (meaning "against the person") arguments try to discredit a claim by attacking its proponents instead of providing a reasoned examination of the claim; we list five varieties of such arguments:

1. *Ad hominem abusive* fallacies attack a person's age, character, family, gender, ethnicity, social or economic status, personality, appearance, dress, behavior, or professional, political, or religious affiliations, and suggest not accepting such a person's views.

2. The fallacy of *guilt by association* (or *poisoning the well*) attempts to repudiate a claim by attacking not the claim's proponent, but the reputation of those with whom they associate or agree.

3. *Tu quoque* ("you too") fallacies attempt to refute a claim by attacking its proponent on the grounds that he or she is a hypocrite, upholds a double standard of conduct, or is inconsistent in enforcing a principle, suggesting that the arguer is unqualified to make the claim, and we should not accept the claim. But, there is an important distinction between a person's words and actions.

4. *Vested interest* fallacies attempt to refute a claim by arguing that its proponents are motivated by the desire to gain something; they suggest that if not for this vested interest, the arguer would hold a different view, and so we should discount their views.

5. *Circumstantial ad hominem* fallacies (similar to vested interest fallacies) attempt to refute a claim by arguing that its proponents endorse two or more conflicting propositions, and so we may disregard one or all of those judgments.

Example 8.1. Classify the following *ad hominem* arguments.

(a) Jones advocates fluoridation of the city water supply.
Jones is a convicted thief.
∴ We should not fluoridate the city water supply.

(b) Jones advocates fluoridation of the city water supply.
Jones spends his free time with known criminals.
∴ We should not fluoridate the city water supply.

(c) Jones believes we should abstain from liquor.
Jones is a habitual drunkard.
∴ We should not abstain from liquor.

(d) Jones supports the fluoridation bill pending in Congress.
He does so because he owns a major fluoridation firm.
∴ We should not support this bill.

(e) Jones says that he abhors all forms of superstition.
Jones also says that breaking a mirror is bad luck.
∴ There probably is something to superstition after all.

(a) An *ad hominem* abusive fallacy. Even if Jones is a convicted thief, this has no bearing on fluoridating the water supply.

(b) A fallacy of guilt by association. Again, the premises are irrelevant to the conclusion.
(c) A *tu quoque* fallacy. Jones's (hypocritical) actions have no bearing on the truth of the conclusion.
(d) A vested interested fallacy. Jones's potential financial stake in the issue at hand has no bearing on the truth of the conclusion.
(e) A circumstantial *ad hominem* fallacy. Jones's claims, whether consistent or not, have no bearing on the conclusion's truth.

All five kinds of *ad hominem* arguments fallaciously attempt to refute a claim by attacking its proponents. In contrast, *straw man* arguments attempt to refute a claim by confusing it with a less plausible claim (the straw man) and then attacking that less plausible claim instead of addressing the original issue. Even a good argument against the less plausible claim is irrelevant to the real issue.

Example 8.2. Evaluate the following argument.

There can be no truth if everything is relative.
∴ Einstein's theory of relativity cannot by true.

The premise is irrelevant to the conclusion, because Einstein's theory does not assert that everything is relative. Rather the claim that "everything is relative" is a straw man attacked by this argument, rather than examining Einstein's actual theory.

Ad baculums (also called *appeals to force* or *appeals to the stick*) fallaciously attempt to establish a conclusion by threat or intimidation.

Example 8.3. Evaluate the following argument.

If you don't vote for me, I'll tell everyone you are a liar.
∴ You ought to vote for me.

The premise is irrelevant to the justification of the conclusion. Coercion, threats, and intimidation may be persuasive at times, but they do not constitute rational argument.

Ad verecundiam fallacies (or *appeals to authority*) occur when we accept (or reject) a claim merely because of the prestige, status, or respect we accord its proponents (or opponents).

Example 8.4. Evaluate the following argument.

> Her teacher says that I should be happy to be an American.
> ∴ I should be happy to be an American.
>
> This argument is an appeal to authority. Without some further evidence that the teacher's statement is correct or justified, the premise is irrelevant to the conclusion.

On many occasions an appeal to authority is either justified or unavoidable. Indeed, in a complex society, where labor is divided and expertise segregated into specialties, much of our knowledge is unavoidably based on appeals to authority. For example, few of us have the required background in mathematics and physics to confirm the equation $E = mc^2$, so it is reasonable to take the word of Einstein and the community of contemporary physicists. Nevertheless, appeals to authority are fallacious if they demand uncritical acceptance of the authority's statements without evidence of the authority's reliability.

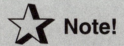 **Note!**

Appeals to authority are not fallacious provided we have good evidence that the authorities have adequate justification for their views.

In contemporary North America, perhaps the most prevalent form of appeal to authority is the *testimonial*, exemplified by celebrities who endorse products, services, or brands of consumer goods. For example, consider the following competing opinions of authorities:

> Leslie Nielsen urges us to buy a new Chrysler.
> ∴ We should buy a new Chrysler.

Kevin Costner urges us to buy a new Ford.
∴ We should but a new Ford.

Both of these arguments are testimonial versions of the fallacy of appeal to authority, exploiting fame or notoriety rather than special knowledge. Now, if Nielsen or Costner urged aspiring thespians to enroll at a particular actor's school, this opinion (as the physics community's endorsement of $E = mc^2$) would gain some relevance since Nielsen and Costner are both successful actors. As a general rule, an appeal to authority is relevant (and hence reasonable) in proportion to the reliability of the authority in the corresponding field.

Ad populum fallacies (or *appeals to the people*) occur when we infer a conclusion merely on the grounds that most people accept it. This fallacy has the form: "X says that P ∴ P" and so is analogous to an appeal to authority, only the X now stands for the opinion of the majority. Appeals to the people often encourage the *bandwagon effect*, relying on peer pressure or social conformity, asking us to join forces with others. Similarly, advertisers often aim their commercial pitch at *cognoscenti*, those allegedly "in the know" on a particular topic.

Example 8.5. Evaluate the following arguments.

   (a) More people drive Chevrolets than any other car.
       ∴ Shouldn't you? (A question indicating you should)
   (b) Discriminating palates prefer wine x.
       ∴ You should drink wine x.

Argument (a) is a bandwagon version of an *ad populum* fallacy; the premise is irrelevant to the conclusion. Argument (b) is a *cognoscenti* version of an *ad populum* fallacy; no reason is given for why you should drink what "discriminating palates" prefer.

Ad misericordiam fallacies (or *appeals to pity*) ask us to excuse or forgive an action on the grounds of extenuating circumstances. An appeal to pity may be either legitimate or fallacious, depending on whether or not the allegedly extenuating circumstances are relevant.

Example 8.6. Evaluate the following argument.

> Officer, you see my baby here was crying for some candy and I took
> her to the candy store before I came back to my car.
> ∴ You shouldn't give me a parking ticket.

This is an *ad misericordiam* fallacy. The arguer appeals to the offi-
cer's pity, but the appeal is of questionable relevance.

Ad ignorantiam fallacies (or *appeals to ignorance*) have the form:
"It has not been proven that *P* ∴ ~*P*". Consider the following examples:

Example 8.7. Evaluate the following arguments.

> (a) No one has ever proven that God does exist.
> ∴ God does not exist.
> (b) No one has ever proven that God does not exist.
> ∴ God does exist.

Both of these arguments are fallacious appeals to ignorance. Nothing about the existence of God follows from our inability to prove
God's existence or nonexistence.

Appeals to ignorance suggest a false dichotomy: either our evidence
for a claim is conclusive or the claim itself is false. However, a claim may
be true even if our evidence for it is inconclusive. In the absence of proof,
the rational approach is to weigh the available evidence, and if the pre-
ponderance of evidence favors one conclusion, to adopt that conclusion
tentatively. If the available evidence is not sufficient to favor a tentative
conclusion, we should suspend judgment.

Ignoratio elenchi fallacies (or *missing the point*) occur when the
premises of an argument warrant a different conclusion than the one the
arguer draws. This is particularly problematic if the "real" conclusion
contradicts or undermines the one actually argued.

Finally, a *red herring* is an extraneous or tangential matter used to
divert attention away from the issue posed by an argument. Typically, a
red herring is irrelevant and contributes nothing to an argument, although
it may mislead its audience into thinking otherwise.

Example 8.8. Evaluate the following argument.

> Some members of the police force may be corrupt, but there
>   are corrupt politicians, plumbers, and even preachers.
> There are also lots of honest cops on the job.
> ∴Let's put police corruption in perspective (the implication being
>   that it's not as bad as it may seem).

The argument attempts to lull its audience into complacency about
the real issue of what to do about corrupt police.

Circular Reasoning

Circular reasoning (also called *petitio principii* or *begging the question*)
occurs when an argument assumes its own conclusion. Such an argument
is always valid (since if the assumptions are all true, the conclusion must
also be true) and is relevant (for what could be more relevant to a conclusion than that conclusion itself?). Furthermore, if all the assumptions
are true, the argument is sound. However, circular reasoning is a fallacy,
for it does not actually *prove* its conclusion.

You Need To Know

Circular reasoning is useless for *proving* its conclusion. If the conclusion is doubtful, so is the corresponding identical assumption and an argument
employing doubtful assumptions has no credibility.

In practice, question-begging arguments are often disguised by restating
the premise in different words than the conclusion, or by keeping one of
the two identical statements implicit.

Example 8.9. Evaluate the following arguments.

  (a) Public nudity is immoral because it is just plain wrong.

(b) Capital punishment is justified. For our country is full of criminals who commit barbarous crimes. And it is perfectly legitimate to punish such inhuman people by putting them to death.

Both arguments beg the question. In (a), the premise "Public nudity is just plain wrong" and the conclusion "Public nudity is immoral" are essentially identical. Similarly, (b) argues for the conclusion that capital punishment is justified by assuming that it is legitimate to put certain criminals to death.

Question-begging epithets are phrases that prejudice discussion and, in a sense, assume the very point at issue. Often they suggest an *ad hominem* abusive attack: "godless communist," "knee-jerk liberal," "Neanderthal conservative," and "inhuman people" as in Example 8.9(b).

Finally, we mention that *complex questions* can be rhetorical tricks somewhat akin to question-begging arguments. For example, the question "Have you stopped beating your spouse?" presupposes an answer to the logically prior question "Have you beaten your spouse?".

Semantic Fallacies

Semantic fallacies occur when the language employed to express an argument has multiple meanings or is excessively vague in ways that interfere with the assessment of the argument's cogency.

Ambiguity (or *equivocation*) is multiplicity of meaning and results from a word or phrase having more than one meaning. This alone does not inhibit understanding, since context usually makes the intended meaning clear. For example, "ball" usually means something different at a sports stadium than in a dance pavilion. Despite context clues, abstract terms such as "right" and "law" are prime candidates for equivocation since we often run their different meanings together. For example, a political right is not the same thing as a legal or moral right— or as *the* political right (i.e., conservatism). Typically, ambiguity generates fallacies when the meaning of an expression shifts during the course of an argument causing a misleading appearance of validity.

Example 8.10. Evaluate the following argument.

> It is silly to fight over mere words.
> Discrimination is just a word.
> ∴ It is silly to fight over discrimination.

This argument commits the fallacy of ambiguity. In the argument, "discrimination" is intended to mean the word "discrimination" itself in the premises, but an action or policy based on prejudice or partiality in the conclusion. The premises may both be true, but they are irrelevant to the conclusion and the argument is invalid.

Ambiguous reasoning frequently requires reference to several different interpretations. No single interpretation can claim to be the argument that is *really* expressed (unless we have conclusive evidence that the arguer intends one). Rather, careful analysis requires attention to all plausible interpretations.

Amphiboly is ambiguity in sentence structure, i.e., ambiguity not traceable to any particular word in a sentence but to how the words are assembled. The occurrence of both universal and existential quantifiers can be one source of amphiboly. For example, the sentence "Some number is greater than any number" has two potential meanings:

1. For any number x, there is some number (not necessarily the same in every case) which is greater than x.
2. There is some number y which is greater than all numbers.

The duality of meaning in this sentence is not traceable to any single ambiguous word, but instead is a product of its overall structure.

Vagueness is indistinctness of meaning, as opposed to multiplicity of meaning. For example, the potential premise "Discriminating palates prefer wine x" is vague. Exactly what is a "discriminating palate" and who has one? Anyone who likes wine x? We cannot determine if vague premises are true, and so we should not accept arguments using them.

Finally, *accent* refers to emphases that generate multiple (and sometimes misleading) interpretations. Newspaper headlines, contractual fine print, commercial giveaways, and deceptive contest entry forms are frequent sources of fallacies of accent.

Example 8.11. Why is the following newspaper headline deceptive?

STUDENTS DEMONSTRATE
New Laser Beam Techniques
Used to Retrieve Coins from Vending Machines

The first line encourages the reader to believe that students are engaged in political protest. The second leads us expect a major scientific breakthrough. The third again forces us to reinterpret the entire message. Drawing conclusions based on the first couple lines would be a fallacy of accent.

Inductive Fallacies

Inductive fallacies occur when the inductive probability of an argument (i.e., the probability of its conclusion given its premises) is low, or at least lower than the arguer thinks it is.

Hasty generalization means fallaciously inferring a conclusion about an entire class of things from inadequate knowledge of some of its members. A hasty statistical generalization often follows from biased, unrepresentative, or inadequate sampling techniques.

Example 8.12. Evaluate the following argument.

Last Monday I wrecked my car.
The Monday before that my furnace broke.
∴ Bad things always happen to me on Mondays.

This is a fallacy of hasty generalization. The inductive probability is extremely low; two bad events on two Mondays are not sufficient evidence to warrant a conclusion about all Mondays.

Faulty analogy is an inductive fallacy associated with analogical reasoning. In analogical reasoning, we assert that object x has certain similarities with object(s) y, and that y has a further property P. We then conclude that x also has property P. For example, we may reason that since rats have many relevant physiological similarities to human beings, and since a certain chemical has been shown to cause cancer in rats, the same chemical is likely to cause cancer in humans. However, the inductive probability of analogical reasoning depends quite sensitively on the degree and relevance of the similarity. If the similarity is slight or not relevant, then a fallacy is likely to result.

Example 8.13. Evaluate the following argument.

> In 1776, the American colonies justly fought for their independence.
> Today, the American Football Alliance is fighting for its independence.
> ∴ The American Football Alliance's cause is just too.

This argument commits the fallacy of a faulty analogy. The disputes resulting in the Revolutionary War included religious liberty, repressive taxation, quartering of troops, and national sovereignty, whereas the (imagined) football alliance may be seeking the right to compete for players, spectators, and television contracts. The analogy is extremely weak since the causes have very little in common, except for the word "independence."

The *gambler's fallacy* is an argument of the form: "x has not occurred recently ∴ x is likely to happen soon." This sort of reasoning is fallacious if x is an event whose occurrences are independent (i.e., if one occurrence of the event does not affect the likelihood of other occurrences). For example, separate occurrences of rain or a favored outcome in a game of chance are usually independent events.

Example 8.14. Evaluate the following arguments.

> (a) On the last ten throws, this fair coin has come up heads.
> ∴ If tossed again, this fair coin should come up tails.
> (b) This clock chimes every hour.
> It has not chimed for about 50 minutes.
> ∴ The clock will chime soon.

Argument (a) is an instance of the gambler's fallacy. For a fair coin, the tosses are independent events and the probability of tails on the next toss is one half regardless of previous results. In contrast, argument (b) is not fallacious since the clock chimings are not independent events as indicated by the first premise.

False cause occurs when the conclusion is a causal claim that is inadequately supported by its premises. This includes confusing a cause with an effect or offering a causal explanation for an event without considering alternatives. Another variant is *post hoc ergo propter hoc* (abbreviated *post hoc*), in which a causal relationship is inferred merely from the temporal proximity of two events.

Example 8.15. Evaluate the following arguments.

(a) Every prophet or messiah is a charismatic leader.
∴ Exercising a talent for leadership is a road to religious inspiration.

(b) The patient became violently ill after eating lunch.
There were no signs of illness prior to eating.
She is in overall good health with a clear medical history.
∴ She was a victim of food poisoning.

Argument (a) is false cause reasoning. Even if the premise is true (which is unclear), this does not make the conclusion probable. A correlation between leadership and inspiration does not mean that leadership causes inspiration. Perhaps religious inspiration leads to leadership, or perhaps there is some other possibility. Argument (b) is a case of *post hoc* reasoning. Although the premises do tell us more than that the illness occurred right after lunch, it is not sufficient to render the conclusion very probable; we need more.

 **Note!**

A *post hoc* argument does not necessarily have a false conclusion. Rather, the evidence in the premise does not *by itself* make the conclusion very probable.

Formal Fallacies

Formal fallacies occur when we misapply a valid rule of inference or follow a "rule" that is invalid. If a formal fallacy is suspected, it is important to determine *both* that the "rule" of inference is invalid (via the methods of formal logic) *and* that the argument itself is invalid by finding a *counterexample* (an actual or logically possible case in which the premises of the argument are true and its conclusion is false).

Example 8.16. Evaluate the following argument.

> If it rains heavily tomorrow, the game will be postponed.
> It will not rain heavily tomorrow.
> ∴ The game will not be postponed.

This argument commits the fallacy of *denying the antecedent*. First, the argument form is invalid $R \rightarrow P, \sim R \therefore \sim P$, as can be verified via truth tables or refutation trees. Second, we show the argument itself is invalid by finding a counterexample. For example, it snows heavily tomorrow and the game is postponed because of the snow. Thus, the argument is invalid.

Example 8.17. Evaluate the following argument.

> If anyone knows what happened, Richard knows.
> No one knows what happened.
> ∴ Richard does not know what happened.

If the premises are true, this argument is valid even though it has the form of denying the antecedent. For if the second premise asserting that no one knows what happened is true, then the conclusion that Richard does not know must also be true.

Important

Having an invalid argument form does *not* automatically mean that an argument commits a formal fallacy; the argument must also be invalid.

Another formal fallacy closely related to denying the antecedent is *affirming the consequent*, which occurs when using the invalid rule of inference: $P \rightarrow Q$, Q ∴ P. For example, "If Smith inherited a fortune, then she is rich. She is rich. ∴ She inherited a fortune." A typical counterexample for this argument is the scenario in which both the premises are true, but Smith made her fortune by creating a successful company. Denying the antecedent and affirming the consequent are similar to, and sometimes confused with, modus tollens and modus ponens, respectively.

Finally, we consider the formal fallacies of *composition* and *division*. The fallacy of composition occurs when we invalidly impute characteristics of one or more parts of a thing to the whole of which they are parts, while the fallacy of division occurs when we invalidly impute characteristics of the whole to the parts.

Example 8.18. Evaluate the following arguments.

(a) Every sentence in this book is well-written.
∴ This book is well-written.
(b) This book is written in English.
∴ Every sentence in this book is in English.

Argument (a) commits the fallacy of composition. For a counterexample, consider a book in which every sentence satisfies grammatic and aesthetic criteria, but the sentences have no relation to one another resulting in incoherence. Argument (b) commits the fallacy of division. For a counterexample, consider a book written in English, but with some sentences written in another language.

Fallacies of False Premises

Arguments with false premises commit the fallacy of false premises. Such arguments are valid, but not sound. One common instance of this type of fallacy is *false dichotomy*, in which the argument makes the false assumption that only one of a number of alternatives holds.

IN THIS CHAPTER:

- ✔ *Statement Strength*
- ✔ *Statistical Syllogism*
- ✔ *Statistical Generalization*
- ✔ *Inductive Generalization and Simple Induction*
- ✔ *Induction by Analogy*
- ✔ *Mill's Methods*
- ✔ *Scientific Theories*
- ✔ *The Probability Calculus*

Statement Strength

In an inductive argument, the conclusion need not follow from the premises as a matter of logical necessity. Rather, in inductive reasoning, we consider the *inductive probability* of the conclusion given the premises. Since the inductive probability of an argument depends on the relative strengths of the premises and conclusion, we begin with a discussion of statement strength.

127

The strength of a statement is determined by what is says; the more a statement says, the *stronger* it is, regardless of its truth value. A *strong* statement is true only under specific circumstances and the world must be just so for it to be true. In contrast, a *weak* statement is true under a wide variety of circumstances and demands little of the world for its truth. The strength of a statement is approximately inversely related to its *a priori* probability (that is, its probability prior to or in the absence of evidence). The stronger a statement is, the less probable it is to be true; the weaker it is, the more probable it is.

The strongest possible statements are those which say so much that they cannot be true; these contradictions are false in every circumstance. In contrast, the weakest possible statements are those which are logically necessary; these tautologies are true under all possible circumstances, and in a sense they say nothing at all.

Example 9.1. Classify the following statements as strong or weak.

(a) Something exists.
(b) Hobbits are humanoid creatures, rarely over a meter tall, with ruddy faces and woolly toes, and live in The Shire.
(c) There are exactly 200 cities with populations over 100,000 in the United States.
(d) It is not true that Knoxville, Tennessee, has exactly 181,379 inhabitants at this very second.

(a) Weak
(b) Strong
(c) Strong
(d) Weak

The weakness of Example 9.1(d) may seem surprising, since it is very specific with respect to time, place, and number. However, all this statement says is that there are *not* exactly this many people at a particular place and time, and so says very little since it allows the population of Knoxville to be anything except 181,379 (and is true even if Knoxville doesn't exist). On the other hand, if we omit the phrase "It is not true that", then this statement becomes quite strong. Similarly, the negation of the weak statement "Something exists" is the strong statement "Nothing at all exists", which says a great deal.

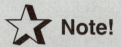 **Note!**

The negation of a weak statement is strong and the negation of a strong statement is weak.

Comparisons of statement strength are not always possible, but we can rank some statements' relative strength via the following:

Rule 1. If statement A deductively implies statement B, but B does not deductively imply A, then A is stronger than B.

Rule 2. If statement A is logically equivalent to statement B (i.e., they imply one another), then A and B are of equal strength.

These rules are immediately relevant to many situations, but not all. Some statements do not imply one another or the difference in strength between statements may be too small to be apparent. In such cases, we cannot order statements with respect to strength.

Example 9.2. Compare the strengths of the following statements.

(a) Some cows are both horned and not horned.
(b) There are cows and buffalo, and all of them are horned.
(c) There are cows, and all of them are horned.
(d) Either some cows are horned or no cows are horned.
(e) Adair admires Adler.
(f) It is not the case that Adair doesn't admire Adler.
(g) Adler is admired by Adair.

The first four statements can be compared and appear in order from weakest to strongest according to Rule 1 and the predicate calculus. The last three statements can also be compared. By the rules of predicate calculus, they are logically equivalent and so they are of equal strength by Rule 2.

The importance of statement strength lies in its relation to inductive probability. The following general rule is fundamental:

Remember

Inductive probability tends to vary directly with the strength of its premises and inversely with the strength of the conclusion.

As we will see, strengthening the premises or weakening the conclusion generally increases inductive probability. However, if we alter the content of statements in a way that disrupts relevance, then this strategy may not increase the argument's inductive probability.

Example 9.3. What effect does increasing the number n have on the following argument form?

> We have observed at least n daisies, all with yellow centers.
> ∴ If we observe another daisy, it will have a yellow center.

The premise gets stronger as the number n gets larger, and so each increase in n increases the inductive probability of the argument.

Example 9.4. Suppose the average height for an adult male in the United States is 5 feet 10 inches. We wish to use this fact as a premise to draw a conclusion about the height of an American man X we have not yet met. Of the following three possible conclusions, which produces the strongest argument?

(a) X is exactly 5 feet 10 inches tall.
(b) X is within a foot of 5 feet 10 inches tall.
(c) X is within an inch of 5 feet 10 inches tall.

As we have discussed, the weaker the conclusion, the higher the inductive probability of the argument. Since (b) is the weakest possible conclusion, this produces the strongest argument and, since (a) is the strongest, this produces the weakest argument.

Statistical Syllogism

Inductive arguments are divisible into two types, according to whether or not they presuppose that the universe, or some aspect of it, appears to be uniform. *Statistical* arguments do not require the presupposition of uniformity; the premise of a statistical argument supports its conclusion via purely mathematical reasons. *Humean* arguments (named in honor of the Scottish philosopher David Hume) require the presupposition of uniformity.

Example 9.5. Classify the following inductive arguments.

    (a) 98 percent of freshmen read beyond the sixth-grade level.
       Dave is a freshman.
       ∴ Dave can read beyond the sixth-grade level.
    (b) All 100 freshmen surveyed can spell "logic".
       ∴ The next freshmen we ask can spell "logic".

Argument (a) is statistical; given the premises, the conclusion is quite likely on statistical grounds. Argument (b) is Humean; the argument is neither deductive nor statistical, but is based on the uniformity of freshmen's spelling abilities.

In the next two sections we consider statistical inductive arguments and then we take up Humean arguments.

You Need to Know

According to the logical interpretation, the inductive probability of a statistical argument is the percentage figure given in the argument divided by 100.

For example, the inductive probability of argument (a) in Example 9.5 is 0.98.

   We first consider *statistical syllogisms*: an inference from statistics concerning a set of individuals to a (probable) conclusion about some member of the set. These are represented in one of the following forms:

> *n* percent of *F* are *G*.   *n* percent of *F* are *G*.
> *x* is *F*.   *x* is *F*.
> ∴*x* is *G*.   ∴*x* is not *G*.

Here, *F* and *G* represent predicates, *x* represents a name, and *n* a number from 0 to 100. For example, argument (a) in Example 9.5 is a statistical syllogism. The inductive probability of such a statistical syllogism (by our logical interpretation) is $n/100$. Typically, the left-hand argument is made when $n > 50$ and the right-hand argument when $n < 50$. However, sometimes the statistics used to draw the conclusion of a statistical syllogism are not numerically precise.

Example 9.6. Evaluate the inductive probability of the following statistical syllogisms.

> (a) Madame P's diagnoses are almost always correct.
>   Madame P says Susan is suffering from a kidney stone.
>   ∴Susan is suffering from a kidney stone.
> (b) Most of what Dana says about his past is false.
>   Dana says he lived in Tahiti with his two wives.
>   ∴Dana did not live in Tahiti with two wives.

The term "almost always" in argument (a) indicates a large percentage and has a high inductive probability. The term "most" in argument (b) means more than half; the inductive probability of this argument is slightly better than 0.5. We cannot provide precise numeric probabilities in light of the premises' vagueness.

 **Note!**

A reasonably high inductive probability is only one of the criteria an argument must satisfy to demonstrate the probable truth of its conclusion. The argument must also have true and relevant premises, and (insofar as possible) must satisfy the requirement of total evidence.

In fact, the premises of a statistical syllogism are automatically relevant by virtue of its form. But they may not be true, and they may not be all that is known with respect to the conclusion. For example, argument (a) of Example 9.6, may have a false first premise; it is an argument from authority, whose strength depends on Madame P's reliability. In contrast, argument (b) of Example 9.6 reasons from the *unreliability* of a person's pronouncements. This is a form of *ad hominem* argument, but the premise addresses the veracity of the person in question (rather than some extraneous attribute), so if the premises are true and there is no suppressed evidence, argument (b) is a reasonably good argument.

Statistical Generalization

Statistical generalization, uses statistics concerning a randomly selected subset of a set of individuals to a (probable) conclusion about the composition of the set as a whole. Statistical generalization is a statistical form of inference, not a Humean form and has general form:

> n percent of s randomly selected F are G.
> ∴ About n percent of all F are G.

The number s indicates the size of the sample, F is a property defining the general population, and G is the property surveyed.

The inductive probability of a statistical generalization is primarily a function of two quantities: the sample size s and the strength of the conclusion. Increasing s strengthens the premise and enhances the argument's inductive probability. In addition, we consider the strength of the argument's conclusion. In the general form of statistical generalization given above, the conclusion is "*About n* percent of F are G." If it said "*Exactly n* percent of F are G," the conclusion would be much too strong and the argument would have an inductive probability close to zero. For our conclusion to be reliable, we must allow a certain margin of error as signified by the term "about".

Example 9.7. Evaluate the inductive probability of the following arguments:

    (a) Fewer than 1 percent of 1,000 ball bearings randomly selected from the 1997 production run failed to meet specs.

∴ Only a small percentage of all the ball bearings produced in the 1997 production run fail to meet specs.
(b) I spoke to three of my friends in the course and they all earned an A.
∴ Virtually everyone in the course earned an A.

The inductive probability of argument (a) is quite high. The size of the sample and the randomness of the sampling process strongly justify the conclusion's generalization. In addition, the generalization involves approximation, as indicated by the phrase "only a small percentage." This weakens the conclusion, adding strength to the argument. In contrast, argument (b) is weak. A sample size of three is too small to justify the generalization expressed by the conclusion. In addition, the sample is "biased" since the friends of the arguer is not a random group of students.

The success of statistical generalization depends crucially on the randomness of the sampling technique. A randomly selected sample is chosen by a method guaranteeing that each of the F's has an equal chance of being sampled, and so each s-membered subset of the F's has an equal chance of being chosen. If a large sample of F's is randomly selected, it is likely (although not certain) that the proportion of G's in the sample approximates the proportion of G's among all the F's. If the sample is not random, then the sampling technique may favor samples with an unusually high or an unusually low number of G's; such samples are said to be *biased*. Applying statistical generalization with a nonrandom sampling technique commits the *fallacy of biased sample*.

Remember

In general, inductive probability is enhanced by increasing the sample size s (which strengthens the premise) and by increasing the margin of error in the conclusion (which weakens the conclusion).

In addition, we note that the percent n has a small effect. If n is very large or very small (near 100 or 0), the argument's inductive probability is higher (other things being equal) than if n is close to 50.

Like all inductive arguments, statistical generalization is vulnerable to suppressed evidence. If two or more random surveys obtain distinctly different results from true premises, then none of them alone constitutes a good argument. There is also a general difficulty with polling human beings: How can we be sure that the respondents are telling the truth? In many cases, there is little motive for dishonesty, but the assumption of truthfulness should not be made uncritically.

A related problem concerns humans' response to how a question is asked. Suppose we wish to survey public opinion on a new piece of legislation proposed by Senator S. How we phrase our question may drastically affect the responses; consider the following:

Do you favor Senator S's government-bloating socialist bill?
Do you favor Senator S's bill on government aid to the poor?
Do you favor Senator S's popular new bill to bring much-needed aid to the victims of poverty in America?

Biased questions can be a problem in polls that are poorly designed or conducted by those with a vested interest in the outcome, and can lead to fallacious statistical generalizations.

Inductive Generalization and Simple Induction

Statistical generalization allows us to arrive at a conclusion concerning an entire population from a premise concerning a *random* sample of that population. However, it is often impossible to obtain a random sample. For example, if the population includes future objects or events, they cannot be included in the sample.

Example 9.8. Evaluate the sample's randomness in the following:

The Bats have won 10 of 20 games so far this season.
∴ The Bats will win about half their games this season.

The conclusion concerns a population (*all* the Bats games this season) including future games, while the sample only includes the games played so far. Therefore, this is not a random sample.

The general form of the argument given in Example 9.8 is called *inductive generalization* and is represented as follows:

> *n* percent of the *s* thus-far-observed *F* are *G*.
> ∴ About *n* percent of all *F* are *G*.

Inductive generalization differs from statistical generalization in that its premise does not claim a random sample. Without randomness, the reasoning cannot be justified by mathematical principles alone, and so inductive generalizations are Humean inferences.

Inductive generalizations are weaker arguments than statistical generalizations, for the uniformity they presuppose is always to some degree uncertain. There is no uniformly accepted way of calculating the inductive probabilities of Humean arguments, so we cannot say exactly how much weaker. However, in other respects, evaluation of inductive generalizations is the same as evaluation of statistical generalizations.

The most notable inductive generalization have the percent *n* = 100. In this case, we have:

> All of the *s* thus-far-observed *F* are *G*.
> ∴ About *n* percent of all *F* are *G*.

This form is the means by which scientific laws are justified. For example, our knowledge that water freezes at +32 degrees Fahrenheit is based (in part) on the fact that all the (many) samples of pure water observed thus far have had freezing point of +32 degrees Fahrenheit.

If the inductive probability of an inductive generalization is not zero, then the probability may be increased by increasing the sample size *s*. Similarly, a smaller population of *F*'s (that is, the weaker the conclusion) yields a greater probability for the argument. The most extreme such weakening of the conclusion is to reduce the population mentioned by the conclusion to one individual. This yields the following argument form, which is called *simple induction* or *induction by enumeration*:

> *n* percent of the *s* thus-far-observed *F* are *G*.
> ∴ If one more *F* is observed, then it will be *G*.

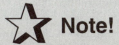

Note!

In general, simple inductions are much stronger than inductive generalizations from the same premises.

Example 9.9. Compare the following arguments based on their inductive probabilities.

    (a) All meteorites contain iron.
       ∴ If a meteorite is observed, it will contain iron.
    (b) All 1,000 meteorites observed thus far contain iron.
       ∴ If another meteorite is observed, it will contain iron.
    (c) Exactly 99 percent of 500 observed meteorites contain iron.
       ∴ If another meteorite is observed, it will contain iron.
    (d) Exactly 99 percent of 500 observed meteorites contain iron.
       ∴ All meteorites contain iron.

The arguments are ranked from highest to lowest inductive probability in the order (a), (b), (c), (d). Argument (a) is a deductive argument. Arguments (b) and (c) are simple inductions ranked by the comparative strengths of their premises. Finally, argument (d) has inductive probability zero; the conclusion is false since at least five observed meteorites do not contain iron.

Induction by Analogy

Another important kind of Humean argument is *argument by analogy*. In an argument by analogy we observe that object x has many properties $F_1, F_2, \ldots, F_n$, in common with some other object y. In addition, we observe that y has some further property G. Thus, we consider it likely (since x and y are analogous in many other respects) that x also has G. The general form of the argument is represented as:

$$F_1 x \ \& \ F_2 x \ \& \ldots \& \ F_n x$$
$$F_1 y \ \& \ F_2 y \ \& \ldots \& \ F_n y$$
$$Gy$$
$$\therefore Gx$$

Example 9.10. Evaluate the following argument:

> Specimen x is a single-stemmed plant with lanceolate leave and five-petaled blue flowers, about 0.4 meters tall.
> Specimen y is a single-stemmed plant with lanceolate leave and five-petaled blue flowers, about 0.4 meters tall.
> Specimen y is a member of the genetian family.
> ∴ Specimen x is a member of the genetian family.

This is a reasonably strong argument by analogy. The argument is Humean since no logical or mathematical principles ensure that similar external appearance, size, and shape yield the same taxonomic type. Rather, the argument assumes a correspondence between these characteristics and taxonomic types. The argument's strength is in part a function of this presupposition's strength.

As with all inductive arguments, analogical arguments may be strengthened by strengthening their premises or by weakening their conclusions. For example, we can raise the inductive probability of the argument in Example 9.10, if we weaken the conclusion to:

> Specimen x is a member of the genetian family or some closely related family.

We can also raise the inductive probability by noting more properties that x and y have in common, thus strengthening the first two premises. For example, we might observe that x and y also produce similar seeds.

Important!

In induction by analogy, some properties count more than others. The strength of a premise depends not only on the number of properties x and y are claimed to have in common, but also the specificity of these properties; more specific similarities yield a stronger argument.

Another consideration is the relevance of the properties $F_1, F_2, \ldots, F_n$ to the property G. If relevance is lacking and the conclusion is strong, the argument's inductive probability will be quite low. Relevance can be difficult to determine, and its role in analogical arguments can be quite problematic. Perhaps the best advice in evaluating analogical reasoning is that common sense should prevail.

Example 9.11. Estimate the inductive probability of the following argument by analogy:

> Person x was born on a Monday, has dark hair, is 5 feet 8 inches tall, and speaks Finnish.
> Person y was born on a Monday, has dark hair, is 5 feet 8 inches tall, and speaks Finnish.
> Person y likes brussels sprouts.
> ∴ Person x likes brussels sprouts.

The inductive probability is low, because the properties $F_1, F_2, F_3,$ and F_4 mentioned in the first two premises are almost surely irrelevant to the property G of liking brussels sprouts.

Analogical considerations can also be combined with simple induction to yield hybrid argument forms. For example, instead of comparing x with just one object y, we may compare it with many different objects, all of which have the properties $F_1, F_2, \ldots, F_n$, and G. This strengthens the argument by showing that G is associated with $F_1, F_2, \ldots, F_n$ in many instances, not just in one.

Finally, analogical arguments, like all inductive arguments, are vulnerable to contrary evidence. If any evidence bearing negatively on the analogy is suppressed, then the argument violates the requirement of total evidence and should be rejected.

Mill's Methods

Often we wish to determine the cause of an observed effect. The first step in doing so is to formulate a list of suspected causes that includes the actual cause. The second is to rule out as many of these suspected causes as possible. If we narrow the list down to one item, we conclude that this item is probably the cause.

> ## ⭐ Note!
>
> The justification of the first step is generally inductive and the eliminative reasoning of the second step is deductive, so the reasoning as a whole is inductive.

We arrive at the list of suspected causes by a process of inductive (often analogical) reasoning. For example, suppose we wish to find the cause of a disease. The disease may resemble some known diseases more than others and we note those to which it is most similar. We conclude (by analogy) that its cause is probably similar to those of the known diseases it resembles, giving us a list of possible causes.

At this point, our investigation is only half finished. To determine which possible cause is actually the cause of the disease, we employ a deductive process designed to eliminate from our list as many of the suspected causes as possible. The type of eliminative process we use depends on the kind of cause we are looking for. There are four different kinds of causes and, corresponding to each, a different method of elimination. The eliminative methods were named and investigated by the nineteenth-century philosopher John Stuart Mill, and are beyond the scope of this text. We define here the different kinds of causes.

First, a condition C is a *necessary cause* for an effect E if E never occurs without C, although perhaps C can occur without E. A given effect may have several necessary causes. For example, fire requires three causally necessary conditions: fuel, oxygen, and heat.

Second, a condition C is a *sufficient cause* for an effect E if it never occurs without E, although perhaps E can occur without C. For example, decapitation is a sufficient cause for (always results in) death among higher animals. However, there are other causes that may result in death, and so a given effect may have several sufficient causes.

Third, a condition C is a *necessary and sufficient cause* of an effect E if E never occurs without C and C never occurs without E. For example, the presence of a massive body is a necessary and sufficient cause for the presence of a gravitational field.

Finally, a *variable quantity B is causally dependent on a second vari-*

able quantity A if a change in *A* always produces a corresponding change in *B*. For example, the apparent brightness *B* of a luminous object varies inversely with the square of the distance of *A* from that object. An effect may be causally dependent on multiple quantities. If our luminous object is a gas flame, its apparent brightness also depends on the amount of fuel and oxygen available, among other factors.

Scientific Theories

The most sophisticated forms of inductive reasoning occur in the justification, or confirmation, of scientific theories. A scientific theory is an account of some natural phenomenon, which in conjunction with further known facts or conjectures (called *auxiliary hypotheses*) enables us to deduce consequences that can be tested by observation.

Scientific theories are justified primarily by their success in making true predictions. By "prediction" we mean a statement about the results of certain tests or observations, not necessarily a statement about the future; even theories about the past make predictions in this sense. For example, a theory about the evolution of dinosaurs will have implications for the sorts of fossils expected in certain geological strata. Such implications are among the theory's predictions. Since a theory's predictions are *deduced* from the theory together with its auxiliary hypotheses, if any prediction is false, then either the theory itself or one or more of the auxiliary hypotheses must be false. In contrast, each observation confirming a theory's predications increases our confidence in the veracity of the theory.

Remember

The reasoning by which scientific theories are refuted is deductive, while the reasoning by which they are confirmed is inductive.

Confirmation of a prediction (or even many predictions) of a theory does not deductively prove the theory is true. Theories together with their auxiliary hypotheses imply more predictions than can be tested, and some untested prediction may still be false. Thus, from a logical point of view, confidence in any scientific theory is never absolute. In practice, it is often held that as more of the predictions entailed by a theory are verified, the theory becomes more *probable*. This principle may be formulated more precisely as follows:

> If E is some initial body of evidence (including auxiliary hypotheses) and C is the additional verification of some of the theory's predictions, the probability of the theory given E and C is higher than the probability of the theory given E alone.

The Probability Calculus

It would be useful to have something analogous to truth tables for inductive reasoning: a procedure that would enable us to calculate probabilities of complex statements from the probabilities of simpler ones and thereby to determine the inductive probabilities of arguments. Unfortunately, no such procedure exists. We cannot always calculate the probability of a statement or the inductive probability of an argument simply from the probabilities of its atomic components.

Yet significant generalizations about probabilistic relationships among statements can be made, shedding a great deal of light on the nature of probability and enabling us to solve some practical problems. The corresponding *probability calculus* is a set of formal rules for expressions of the form "$P(A)$", meaning "the probability of A." For applications of the probability calculus to logic, A generally denotes a proposition, and sometimes an event. The operator P can be interpreted in a variety of ways. The oldest and simplest concept of probability is the *classical interpretation*. Probabilities are defined only when a situation has a finite nonzero number of equally likely possible outcomes. In this case, the probability of A is defined as the ratio of the number of possible outcomes in which A occurs to the total number of possible outcomes:

$$P(A) = \frac{\text{Number of possible outcomes in which } A \text{ occurs}}{\text{Total number of possible outcomes}}$$

Example 9.12. Consider a situation in which a fair die is tossed once. here are six equally likely possible outcomes: the die shows a one, denoted A_1, the die shows a two, denoted A_2, . . . , and the die shows a six, denoted A_6. Calculate the following probabilities:

(a) $P(A_1)$
(b) $P(\sim A_1)$
(c) $P(A_1 \vee A_3)$
(d) $P(A_1 \,\&\, A_3)$

(a) 1/6
(b) $1 - 1/6 = 5/6$
(c) $2/6 = 1/3$
(d) 0/6 (since the one die tossed cannot show two numbers)

As Example 9.12 illustrates, logical properties or relations among events or propositions affect the computation of the probability of complex outcomes. For example, a *contradictory* (truth-functionally inconsistent) event, such as $A_1 \,\&\, \sim A_1$, can never occur and so has probability zero. Events that cannot occur simultaneously are said to be *mutually exclusive* and their conjunction has a probability of zero.

The probability calculus consists of the following three axioms (basic principles), together with their deductive consequences:

Axiom 1. $P(A) \geq 0$.
Axiom 2. If A is tautologous, $P(A) = 1$.
Axiom 3, If A and B are mutually exclusive, $P(A \vee B) = P(A) + P(B)$.

We consider some theorems, or deductive consequences, of our axioms. These theorems enhance our understanding of probability and provide a basis for practical applications of the probability calculus.

Example 9.13. Prove: $P(\sim A) = 1 - P(A)$.

Since $A \vee \sim A$ is tautologous, by Axiom 2, we have $P(A \vee \sim A) = 1$. And since A and $\sim A$ are mutually exclusive, by Axiom 3, $P(A \vee \sim A) = P(A) + P(\sim A)$, and so $P(\sim A) = 1 - P(A)$.

Example 9.14. Prove: If A is contradictory, $P(A) = 0$.

Suppose A is contradictory. Then $\sim A$ is tautologous, since negation changes every T to F in a truth table. Hence, by Axiom 2, $P(\sim A) = 1$ and so, by Example 10.2, $P(A) = 0$.

Proceeding in a similar fashion to the above, we obtain the following identities:

> If A and B are truth-functionally equivalent, then $P(A) = P(B)$.
> $0 \leq P(A) \leq 1$
> $P(A \vee B) = P(A) + P(B) - P(A \,\&\, B)$
> $P(A \,\&\, B) \leq P(A)$ and $P(A \,\&\, B) \leq P(B)$
> $P(A \vee B) \geq P(A)$ and $P(A \vee B) \geq P(B)$
> If $P(A) = P(B) = 0$, then $P(A \vee B) = 0$.
> If $P(A) = 1$, then $P(A \,\&\, B) = P(B)$.
> If A is a truth-functional consequence of B, $P(A \,\&\, B) = P(B)$.

Index